Enhancing Army
S&T

Enhancing Army S&T

Lessons from *Project Hindsight Revisited*

By Richard Chait
John Lyons
Duncan Long *and*
Albert Sciarretta

PUBLISHED BY THE CENTER FOR
TECHNOLOGY AND NATIONAL
SECURITY POLICY AT THE
NATIONAL DEFENSE UNIVERSITY
WASHINGTON, DC
2007

The views expressed in this report are those of the authors and do not reflect the official policy or position of the National Defense University, the Department of Defense, or the U.S. Government. All information and sources for this paper were drawn from unclassified materials.

Portions of this book may be quoted or reprinted without permission, provided that a standard source credit line is included.

This book was published by the National Defense University Center for Technology and National Security Policy, Fort Lesley J. McNair, Washington, DC.

CTNSP publications are available online at http://www ndu.edu/ctnsp/publications.html.

Contents

About the Authors

Richard Chait is a Distinguished Research Fellow at the Center for Technology and National Security Policy (CTNSP) at the National Defense University. He was previously Chief Scientist, Army Material Command, and Director, Army Research and Laboratory Management. Dr. Chait received his PhD in Solid State Science from Syracuse University and a BS degree from Rensselaer Polytechnic Institute.

John W. Lyons is a Distinguished Research Fellow at CTNSP. He was previously director of the Army Research Laboratory and director of the National Institute of Standards and Technology. Dr. Lyons received his PhD from Washington University. He holds a BA from Harvard.

Duncan Long is a Research Associate at CTNSP. He holds a Master of International Affairs degree from the School of International and Public Affairs, Columbia University, and a BA from Stanford University.

Albert A. Sciarretta is president of CNS Technologies, Inc. He is a retired Army armor officer, whose service included operational assignments, instructing at the U.S. Military Academy, acting as technology officer on armored vehicle task forces, and serving as Assistant to the Chief Scientist, Army Material Command. He has two M.S. degrees—Operations Research and Mechanical Engineering—from Stanford University and a B.S. degree from the U.S. Military Academy.

Acknowledgments

The *Project Hindsight Revisited* series of studies and this book would not have been possible without the contributions of a great many people. The willing assistance of the many technologists, managers, and Army officers whom we interviewed and with whom we corresponded was absolutely essential to the undertaking. We have done our humble best to acknowledge in our foregoing reports those who assisted us in our research. Here, we would like to re-emphasize our gratitude for the efforts of Dr. Thomas Killion, the Army S&T Executive, who saw the benefit of looking back at the development of key Army weapons systems for insights that would be helpful in managing S&T in the current and future environment. We would also like to express our appreciation to our colleagues at the Center for Technology and National Security Policy, especially to CTNSP's director, Dr. Hans Binnendijk, who was unstinting in his support. Any errors or shortcomings are, of course, solely the authors' responsibility.

Foreword

In July 1965, the Director, Defense Research and Engineering (DDR&E) provided guidance to the Assistant Secretaries for Research and Development of the military departments to conduct *Project Hindsight*. The two main objectives of the project were to identify those management factors that are important in ensuring that research and technology programs will be productive and program results will be utilized, and to measure the overall increase in cost-effectiveness in the current generation of weapons systems compared to that of their predecessors (when such can be identified) that is assignable to any part of the DOD investment in research in science and technology (S&T). The *Project Hindsight* group made an historical assessment of the research and development "events" that contributed to the development of 20 major weapons systems. The results were published in October 1969. The assessments documented the key role that defense S&T played in enabling the generation of weapons systems fielded in the 1960s and, more specifically, the work defense laboratories and centers, as well as industry, performed in realizing those capabilities.

With this in mind, in 2004 I asked the Center for Technology and National Security Policy (CTNSP) at the National Defense University to conduct an in-depth, follow-on study on Army weapons systems; I informally named it "Project Hindsight Revisited." The study's goal was similar to the work DDR&E had charted in the 1960s: to examine a number of major weapons systems to assess the role of S&T in providing their key capabilities. In the 1960s, as today, reasonable questions could be asked about the relevance and value of any S&T investment—within DOD, across the Federal government and in the commercial sectors. As found in the original study and reinforced in this one, the majority of technology investments that contribute to weapons systems capability are in fact traceable to DOD.

At the time of the original study there was a growing need to understand the factors contributing to effective transition of technology into systems. Then, as today, S&T funding was a resource area under challenge to show "payoff," as results of research were primarily forecasts about future success. The *Hindsight* project and this revisit validated a major factor influencing the effective transition of technology to systems: close collaboration between the technologists and the acquisition community is paramount to achieving success. Processes can facilitate that linkage, but the result really depends on the effectiveness of communication among the people who operate within the processes.

The *Hindsight Revisited* study posed, then tried to answer, several of the most relevant—perhaps enduring—questions on assessing the contribution of S&T to enabling major warfighting systems. First, are defense S&T investments as critical today as in the 1960s, or is more commercial technology relied upon to enable our current systems? Second, has the role of our laboratories and centers changed significantly in the past 40 years? Are there lessons to be learned from recent case studies of technology transition to better shape our S&T program and its infrastructure for the future?

The *Hindsight Revisited* studies covered four major Army weapons systems: the Abrams tank, the Apache helicopter, and the Stinger and Javelin missiles. This book highlights and elaborates on several of the many illuminating insights that may be gleaned from these reports on the S&T efforts that contributed to these programs. It may not be surprising to learn from the study that some of history's lessons remain valid, and that today's environment requires its own innovative solutions. One major difference in technology development in the past 20 years, as the information age has matured, is the exploitation of, and in fact reliance on, advanced modeling and simulation to speed technology development.

This book also confirms the previous assessment that maintaining robust Army S&T investment is critical to speeding future capabilities to soldiers. The predominant driver in achieving success has two dimensions—sustaining commitment of resources to the people who "invent change," and close coupling to rapidly evolving warfighting concepts. To maintain the Army's technology edge, nothing is more important than having a cadre of Army scientists and engineers who are engaged with the combat development community to understand what the Army needs and recognize where technology can make a significant difference for our soldiers. It is these scientists and engineers, and our industry partners, whose patience, persistence, and vision we have viewed through the lens of history in this study.

Thomas H. Killion
Deputy Assistant Secretary of the Army
(Research & Technology)
/Chief Scientist

Executive Summary

This book draws on a series of studies known as Project *Hindsight Revisited* conducted by the authors at the National Defense University from 2004–2006. The *Hindsight Revisited* studies examined, in three reports, the development of four current U.S. Army weapons systems: the Abrams main battle tank, the Apache attack helicopter, the Stinger anti-aircraft missile, and the Javelin anti-tank missile. In exploring how these weapons systems were taken from conceptual design to full-scale production, the studies brought to light crucial factors in their successful development. This book pursues significant implications of the studies' findings, with the intention that this analysis and commentary will help the Army's science and technology (S&T) leadership manage the Army S&T portfolio today and tomorrow.

The *Hindsight Revisited* studies addressed the development of the weapons systems in question in terms of Critical Technology Events (CTEs). CTEs, as further explained in Chapters I and II, were those actions and advances that were vital to the capabilities with which a system was ultimately endowed. As such, they provided a way to focus on those factors that were crucial to success, including who performed the development work, who financed it, and what management practices were employed. These CTEs were established after many hours of discussion and extensive correspondence with scientists, engineers, technicians, and managers who were directly involved with the systems' developments.

All told, the *Hindsight Revisited* studies found 135 CTEs for the four systems. Despite the obvious differences in the development stories of such disparate systems as a tank, a helicopter, and two missiles, our research showed some common ingredients of success. Success depended both on having the S&T resources available to conduct the work and on the ability of the parties, principally the Army laboratories and industry, to team and work together.

The highly competent technical staffs and special-purpose equipment and facilities needed for all four systems were available because of steady, prolonged investment in S&T resources. The Army laboratories provided many of the CTEs and were involved in most of them. The long history of work in weapons-related technologies and the attendant availability of unique facilities (ballistic ranges, simulation capabilities, etc.) was a major factor contributing to this finding. Similarly, industry had amassed facilities and expertise, and provided more CTEs in technical areas, such as engines and transmissions, where there had been long-standing demand, both civilian and military, for private-sector expertise.

The collaborative use of these assets was vital to the successful development of the *Hindsight Revisited* systems. This collaboration took the shape, first and foremost, of work conducted jointly by the Army's in-house laboratories and industry. Around 20 percent of the CTEs noted in the studies were the result of major contributions by both of these parties. Collaboration, though, was also important in other facets of development. The Army laboratories collaborated with other government laboratories, leveraging their expertise and facilities. The in-house laboratories also collaborated with the user community (i.e., the warfighters) to ensure that the weapons systems would provide the needed capabilities. Throughout, the Program Manager's

(PM) offices facilitated this collaboration, providing oversight and performing the vital function of integrating the various contributions.

In addition to these findings on resourcing and management of Army S&T, the *Hindsight Revisited* studies made an important finding pertaining to the systems' technical content. The studies revealed common technical elements among the systems. Though the systems had quite different capabilities, certain technologies were important to all of them. Modeling and simulation (M&S), ballistics, materials sciences, and communications technologies are among the chief examples.

This book explores the future implications of two important aspects of the *Hindsight Revisited* findings: the common technologies necessary to the successful development of the *Hindsight Revisited* systems and the importance of the Army's in-house laboratories to weapons development.

Chapter III discusses six families of technologies that were used in the development of two or more of the *Hindsight Revisited* systems. Three enabled systems development: M&S, high performance computing (HPC), and human-systems integration (HSI). Three gave rise to fielded capabilities: gun technology, night vision technology, and power-source technology. As suggested by their prevalence in the *Hindsight Revisited* studies, these six technologies all offer high leverage S&T investment possibilities—advances in any one of them is likely to have pay-off in more than one, and possibly in many, future systems. In some cases, this book focuses on specific, illustrative targets for investment. In the broad area of power-source technologies, for instance, soldier-portable battery technology advances are particularly important.

Chapter IV focuses on another critically important finding of the *Hindsight Revisited* reports, the importance of the Army's in-house laboratories to successful technology development. The in-house laboratories generated 40 percent of the CTEs we noted, and shared responsibility with industry on 20 percent of the others. The laboratories have suffered in recent years from significant funding cuts and have faced calls for their responsibilities to be turned over to the private sector. Though adjustment in the laboratories' roles and practices to capitalize on private-sector expertise to greatest extent possible may be necessary, it seems clear that in the main in-house laboratories will continue to be absolutely essential to the successful development of Army technologies.

Chapter IV, then, addresses the question of how best to strengthen the Army's laboratories to meet the Nation's future needs. We touch on a variety of challenges, including the diminished attention paid to basic research and constantly varying budgets. Principally, though, we focus on one of the Army laboratories' main problems: attracting and retaining talented personnel. Capital investments can compensate for equipment and facility shortfalls, but a strong staff, the heart of any laboratory, cannot be easily assembled. Moreover, technical competency and adept management is vital to continuing the strong collaboration and teaming with industry that was so much a part of the successful development of the four *Hindsight Revisited* systems. With this in mind, we describe several studies and legislative actions that have addressed these personnel management challenges, from the Civil Service Reform Act of 1978, which authorized a laboratory demonstration program with simplified classification and pay systems, to the 2004 National Security Personnel System. We find that giving laboratory managers authorities that would improve the laboratories' personnel posture, including the power to make the kind of local hiring, classification, and salary decisions needed to attract top talent, is of critical importance in maintaining the strength of the Army laboratories.

The book concludes in Chapter V with a charge to the Army S&T leadership and to the Defense Department to preserve the elements of technical development success found in the *Hindsight Revisited* reports. Collaboration with industry, with other DOD partners, with academia—this has been, is, and will continue to be vital to Army S&T success. The Army has benefited, is benefiting, and will continue to benefit from investment in critical technologies that serve multiple weapons systems, in basic research, and in unique equipment and facilities. The Army laboratories have been, are, and will continue to be crucial to technical success.

Enhancing Army S&T

Chapter 1

Introduction

T he urge to maintain military superiority over potential adversaries has long been a driver of technological advancement. This interplay between defense strength and technology, so evident in the nature of America's military power, has for decades prompted U.S. defense planners to engage in technology forecasting. Analysis of emerging technologies was, and is, vital to making wise defense investments.

While it is important to assess the needs and challenges of the future, understanding past military technological successes can be equally important to Army S&T investment and management. By studying past technology development for weapons systems, one can see what factors were important for success and apply these lessons to the management of S&T[1] for future systems. This is an especially valuable exercise now, because in recent years there has been mounting pressure to transfer much of the execution of technical work away from the military's in-house S&T laboratories to the private sector. Whatever the merits of such a move, it represents a significant change from past practices. It would be unwise to undertake any

---

[1] In this report, we use the term S&T to denote what the Army refers to as Basic Research, Applied Research, and Advanced Technology. These are also referred to by their budget codes: 6.1, 6.2, and 6.3, respectively.

fundamental shifts without first understanding just what was successful about the way the Army S&T program has done business in past years.

This book draws on a series of studies known as *Project Hindsight Revisited* conducted by the authors at the National Defense University (NDU) from 2004–2006. The *Project Hindsight Revisited* studies examined, in three reports, the development of four current weapons systems of the U.S. Army: the Abrams main battle tank,[2] the Apache attack helicopter,[3] the Stinger anti-aircraft missile,[4] and the Javelin anti-tank missile. In exploring how these weapons systems were taken from conceptual design to full scale production, the studies brought to light crucial factors in their successful development. This book will pursue significant implications of the studies' findings. In exploring these findings, we hope to make a contribution to answering a question that is of the utmost importance to the Army leadership: how are S&T resources best used to advance the state-of-the-art capabilities of U.S. Army weapons systems?

The book begins by discussing the findings of *Project Hindsight Revisited*. That series of studies focused on Critical Technology Events (CTEs), the significance of

---

[2] Richard Chait, John Lyons, and Duncan Long, "Critical Technology Events in the Development of the Abrams Tank: *Project Hindsight Revisited*," *Defense and Technology Paper 22* (Washington, DC: Center for Technology and National Security Policy, National Defense University, December 2005).

[3] Chait, Lyons, and Long, "Critical Technology Events in the Development of the Apache Helicopter: *Project Hindsight Revisited*," *Defense and Technology Paper 26* (Washington, DC: Center for Technology and National Security Policy, National Defense University, February 2006).

[4] Lyons, Long, and Chait, "Critical Technology Events in the Development of the Stinger and the Javelin Missiles: *Project Hindsight Revisited*," *Defense and Technology Paper 33* (Washington, DC: Center for Technology and National Security Policy, National Defense University, July 2006).

which are explained below. We used CTEs as a way to focus on the factors that gave rise to a successful weapons system. These factors, which are discussed along with other findings in Chapter II, include where the technical work was performed, the source of funding, management style, and teaming among Army laboratories and private industry. Chapter II pays particular attention to the sources of the CTEs. We conclude the chapter by comparing the findings of *Hindsight Revisited* to the results of an earlier DOD study, the original *Project Hindsight*.

Chapter III expands on an important *Hindsight Revisited* finding. We discovered that certain families of technologies played an important role in the development of all four systems. We discuss therein some of those technologies, focusing on those areas in which S&T investment is likely to pay off in capability improvements for multiple weapons systems. This includes both technologies that enable weapons development, such as modeling and simulation (M&S), and technologies in fielded systems, such as night vision. We also touch briefly on a group of technologies that will impact the entire future force: those supporting network-centric operations.

Chapter IV draws on another important *Hindsight Revisited* finding: Army laboratories were a fundamental part of the successful development of the studied systems. Despite the obvious differences in the development stories of such disparate systems as a tank, a helicopter, and two missiles, our research showed some common ingredients of success in each case: good Army laboratory facilities and skilled staff. Chapter IV focuses in particular on this latter ingredient; in addition to a variety of other challenges, the Army laboratories face significant problems in attracting and retaining talented personnel.

Finally, with some of the key factors that led to Army S&T success in hand, we conclude in Chapter V by looking ahead to the future of Army S&T. Drawing on the findings of *Hindsight Revisited* and on the discussion in this book, we make recommendations for maintaining a robust Army S&T program. With these recommendations, the technological prowess that has under-girded American military might in the past decades will be maintained and enhanced.

CTEs, Innovation, and the *Hindsight Revisited* Approach

The CTE was the lynchpin of the *Hindsight Revisited* studies. CTEs are ideas, concepts, models, and analyses, including key technical and managerial decisions, that had a major impact on the development of a specific weapons system. CTEs can occur at any point in the system's life cycle, from basic research, to advanced development, to testing and evaluation, to product improvements. CTEs can even relate to concepts that were developed but ultimately not incorporated into the weapons system. Also, they can originate in many places: the Army's in-house laboratories, the private sector, academia, and the S&T programs of our international partners. The CTEs, as hallmarks of technical advances, gave us a way to focus our attention on the important factors for success. We did not attempt to capture every single technical development in a given system or to discuss its development in exhaustive technical detail.[5]

---

[5] Some technologies not discussed nor included in the CTE count turned out to be relatively significant advances. An example is the Explosive Reactive Armor (ERA), which was considered for the Abrams. Though ERA was not initially fielded with the tank, it was developed as an add-on kit and has been fielded today. Also worthy of note is aluminum oxynitride (ALON^{TM}), a transparent ceramic material with high hardness and excellent ballistic properties. This material was patented by the Army in the mid-1970s and, with industry effectively addressing production and development issues, has become a leading candidate for transparent armor in next generation systems.

Given the central importance of the CTE to our reports, it is worthwhile to explain the concept in greater detail and place it in the broader context of technical innovation. A good framework for understanding the innovation process and the genesis of operationally useful advances is provided by a recent NDU report, "The S&T Innovation Conundrum," by Coffey, Dahlburg, and Zimet.[6] As illustrated in Figure 1, the authors hold that S&T innovation contains two distinct phases, "prospecting" and "mining." Prospecting is early work that provides a fundamental scientific basis for later research. Mining is later work to develop specific systems. The prospecting phase is often not focused on a particular outcome for a particular system. The mining phase draws on knowledge gained in the prospecting phase to yield useful capabilities.

---

[6] Timothy Coffey, Jill Dahlburg, and Elihu Zimet, "The S&T Innovation Conundrum," *Defense and Technology Paper 17* (Washington, DC: Center for Technology and National Security Policy, National Defense University, August 2005).

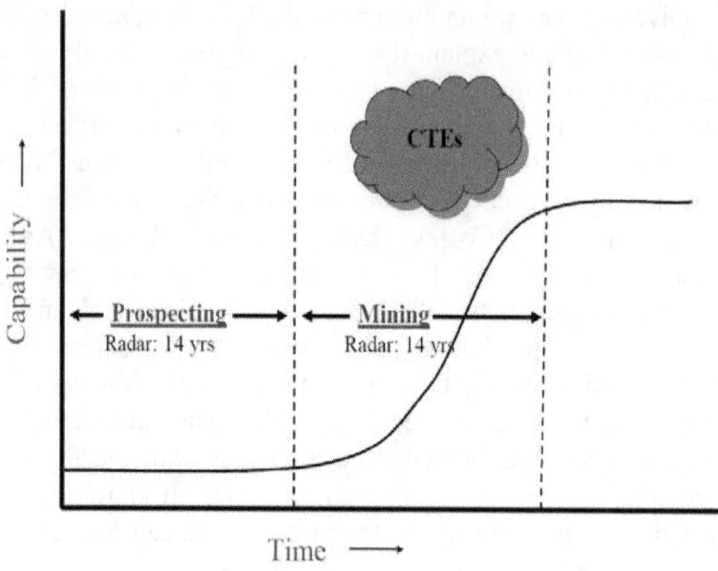

Figure 1. Prospecting and mining phases of
development of radar.[7]

The majority of our CTEs occurred in the mining
phase—the later stages of the basic research-to-engineering
development continuum. This is largely because we elected
to count as CTEs only those things that bore on significant
improvements over the predecessor systems. The Abrams
was compared to the M–60 Patton tank, the Apache to the
AH–1 Cobra, the Stinger to the Redeye, and the Javelin to
the Dragon. We could have broadened the time horizon of
our examination considerably and addressed more basic
research. For example, to discuss infrared vision systems
for the Abrams tank we could have looked back to the
development of the quantum theory in the early 20th
century and the subsequent development of solid state
physics in mid-century and found CTEs all along the way.
Doing so would not have shed much light on what was a

---

[7] Drawn from Coffey, Dahlburg, and Zimet.

critical technology in development of the particular weapons systems and would have shifted focus from the specific factors that have produced great improvements in the performance of the weapons systems studied in comparison to those they replaced.

It is important to note that basic research need not be as non-specific and general as the above reference to quantum theory would suggest. It can be problem-driven as well as curiosity-driven. Though DOD defines basic research as the scientific study of phenomena not related to any one specific weapons system (that is to say, knowledge-based or curiosity-driven),[8] and while there are many examples of curiosity-driven research in DOD (most clearly in the work sponsored at universities), a majority of DOD basic research is problem-driven. It is also vital to emphasize the importance of basic research suggested by the prospecting/mining concept. Basic research is needed to lay the foundations for later advances; someone must do the prospecting if there is to be any mining.

These two points—the need for basic research to tackle specific problems and the need for basic research as a foundation for future advances—are both illustrated by Figure 2, which charts the evolution of the technology associated with the M829 series of long-rod penetrators. The figure shows the progression from a mix of curiosity- and problem-driven basic research through applied research and advanced development leading to one version of the M829 series—the M829A1. The work done in the research phase was the basis for many developments for the long-

---

[8] This definition, however, is not followed in practice. Nor should it be, according to a recent National Academies study, *Assessment of Department of Defense Basic Research* (Washington, DC: National Research Council, 2005). This NRC report recommends a concept of basic research that includes both curiosity-driven work (not addressing a particular system) and problem-driven work (applied to a particular problem or problems in developing a system).

rod penetrator, such as the use of composite materials for the sabot of the M829A3 round (CTE No. 16 in the Abrams tank report). These efforts, which demonstrate the importance of basic research to providing a stream of new ideas on which to base future CTEs, involved contributions from various entities, including the Army laboratories, industry, Department of Energy (DOE) laboratories, and academia.

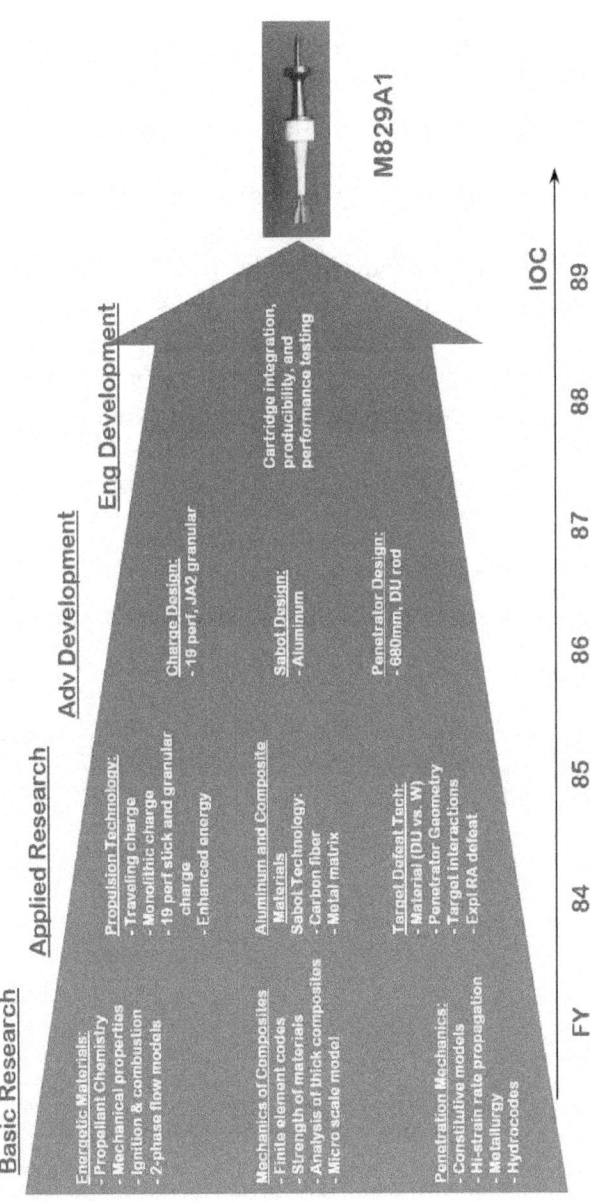

Figure 2. From basic research to the final product for the M829A1 (Figure from the Office of the Secretary of the Army).

We embarked on the *Hindsight Revisited* studies, then, with the CTE as a clear guide to our efforts. We focused these efforts on interviews and correspondence with individuals who were personally involved in the development of the Abrams, the Apache, the Stinger, and the Javelin. In many cases the events described occurred as many as 40 years ago, making it somewhat difficult to reconstruct the details in every case, yet this rich store of first-hand knowledge was indispensable, given the spotty availability of relevant documents and books. With the assistance of these knowledgeable professionals, we were able to map out the CTEs for each of the systems. These CTEs and the studies' broader findings are discussed in the following chapter.

Chapter 2
Hindsight Revisited
Findings and Analysis

T he *Hindsight Revisited* reports revealed some
valuable insights into the factors that led to some of
the Army S&T program's most important successes.
This chapter reviews the studies' results, covering the
number and source of the CTEs. Our study of the
development of the Abrams, the Apache, the Stinger, and
the Javelin yielded 135 total CTEs. The CTEs for each
system are discussed briefly below (they are listed in full in
Appendix C). We then move on to three areas of analysis.
The first is a look at the sources of the CTEs, in which we
account for what kind of organization—government
laboratory, private industry, or other entity—was
responsible for a given advance. The second is a discussion
of the qualitative findings we obtained after reviewing the
systems collectively. We then conclude our analysis of the
Hindsight Revisited reports by comparing our studies'
findings to those of the original *Project Hindsight*, a DOD-
wide review completed in 1969.

Hindsight Revisited CTEs

The Abrams Tank

We identified 55 CTEs in the development of the Abrams tank. The majority of these, 31 in total, were related to armor and armaments. There were nine CTEs related to the power train and 15 related to vehicle electronics, fire control, and communications systems. As indicated by our definition of a CTE, not all of these CTEs were technical developments per se. Some, like the choice to use a gas turbine engine rather than a diesel engine in the tank, in our opinion were management decisions. Others were related to laboratory capabilities, like the use of advanced M&S tools to conduct analysis for simulated live fire tests.

The Apache Helicopter

We identified 44 CTEs in the development of the Apache helicopter. Ten of these had to do with the power system. Fourteen CTEs were related to crew protection, an area on which the Army placed special emphasis, and other structural considerations. We found 19 CTEs dealing with the helicopter's avionics, fire control system, and weapons suite, and three CTEs related to M&S work and other enabling technology factors.

The Stinger and Javelin Missiles

We identified 36 CTEs for the Stinger and the Javelin missiles. For both systems, the seeker and the guidance system were important technical areas. We found six CTEs for the Stinger's seeker and three for guidance and control systems. For the Javelin's seeker we found four CTEs, guidance and control had two CTEs, and the command launch unit had six CTEs. For the propulsion and warhead components we found for the Stinger, three CTEs and for the Javelin, eight CTEs. In M&S we list two CTEs for

Stinger and one for Javelin. The Javelin also had two CTEs that were based on management decisions.

Assessing the Contributions of the Participants

The count and distribution of the CTEs for the four systems give some indication of technical areas that produced significant increases in capability.[9] CTEs, though, were a means to an end: they were an instrument to identify important technical advances so that we could examine what factors gave rise to key achievements. One indisputably important factor in the creation of a CTE is who or what entity was responsible for the crucial breakthrough. We therefore analyzed each CTE to assess who, in our judgment, was the principal source for the event.

We established four categories of work performers: *government technical facilities,* to include all government/in-house laboratories; *private industry,* to include prime contractors and sub-contractors; *joint government/industry,* for any CTE in which both government laboratories and industry played a substantial role; and *other,* for the accomplishments of universities and foreign governments and for CTEs that were government management decisions rather than technical advances. It is likely that no CTE represented the exclusive contribution of one of these four groups, but except in those cases where multiple contributions were substantial enough to warrant the label "joint," we categorized the CTE based on where the clear majority of work was performed.

---

[9] Please note, though, that the CTE count is an imperfect measure of the value of the work done in a given technical area. A single CTE related to the Javelin's seeker, for instance, could add more to the missile's capability than several CTEs related to propulsion.

Table 1 below contains the results of this CTE source analysis. For the Abrams, the in-house laboratories led with just over half of the total CTEs. If we add to this the in-house role in joint work with industry, a little over 60 percent of the work involved the in-house laboratories. Around a quarter came exclusively from outside the government. The figures for the Abrams are not surprising given the preeminent role of the Army laboratories in the fields of armor and armaments. For the Apache, the government laboratories played a somewhat lesser but still crucial role. Industry played a greater role than for the Abrams. For both the Stinger and the Javelin, industry made major contributions; the in-house laboratories played less of an independent role than they had for the Abrams and the Apache, but they were important players in joint developmental work.

System	Gov't/In-House Labs	Joint	Industry	Other	CTE Total
Abrams	55%	18%	13%	14%	55
Apache	43%	20%	30%	7%	44
Stinger and Javelin	17%	25%	50%	8%	36

Table 1. CTE Source Analysis.

This breakdown reinforced for us the vital role played by the government laboratories in developing the necessary technology and the importance of a strong collaborative relationship between the government laboratories, industry, and the Program Managers' (PM) offices.

Overarching Findings from *Hindsight Revisited*

We drew findings and conclusions on each of the specific weapons systems in our four *Hindsight* reports. While each of the weapons systems is unique, their developments had many common elements. We have compiled these common elements as overarching findings below.

1. Funding was almost entirely from the Department of Defense.

The work was spread over many different laboratories in different DOD agencies. The Defense Advanced Research Projects Agency (DARPA) provided some funding, especially for the Javelin missile, but the majority of the money came from Army S&T accounts. In some cases work by contractors was done using Independent Research and Development funds, but these too came (indirectly) from DOD. In a few areas, investments by our international partners produced useful additions to the systems. For example, in the case of the Abrams, the United Kingdom performed work on composite armor and development of the 120mm gun was done by Germany.

2. In-house laboratories and industry were the primary sources of CTEs.

Although CTEs came from a variety of sources—in-house laboratories, industry, academia, international partners—the in-house laboratories in particular played critical roles in the development of each system. Not only did the laboratories contribute many of the CTEs themselves, but they also were partners in important collaborations with industry. The in-house laboratories played another critical role as evaluators of performance and as technical consultants to the contractors and the PM offices. They were able to do this because they had been able to maintain continuity of staff expertise in the important technical disciplines and had the necessary equipment and facilities. These functions helped ensure that the Army was a "smart buyer" as well as an able collaborator.

Industry's technical staffs also made a great many very important technical advances. They were also responsible for the design, integration, and production of the components and final systems. In each case they were successful because they had a trained, experienced workforce and established facilities.

3. Collaboration among the several participants was critical for success.

A common characteristic of the CTEs was the ability of the participants to work smoothly across organizational lines. Cooperation between the in-house laboratories and private industry was especially fruitful: note that 20 percent of the CTEs were joint efforts. Collaboration with the private sector has long been a hallmark of Army S&T, much of it at the level of the working technologists "at the bench." The Army laboratories also worked closely with other government organizations and with academia. The collaboration with the National Aeronautics and Space Administration (NASA) on the Apache, for instance, was critical to the system's development.

4. Systems integration was key to the technology transition process.

For a weapons system to succeed each subsystem must be physically and functionally compatible with all the others. Integration was handled mostly by industry and overseen by the PM. We did not set out in our reports to study systems integration per se, but the topic kept arising. It became clear that system integration is as important as the development of the component technologies. Integration decisions usually require tradeoffs between cost, size and weight, and functionality. Making good trades requires intimate knowledge of the needs of the ultimate user, the soldier in the field. Integration is based on detailed technical knowledge and skills in managing people and organizations. In the systems that we studied, integration was achieved by very close working relationships—sometimes continuing collaborations— among the participants. We believe these relationships were vital to the success of the programs.

Recently, the DOD has begun to use Lead Systems Integrators (LSIs) from industry to handle very large, complex systems programs such as the Future Combat Systems (FCS) program. The LSIs appear to play many of the roles heretofore associated with the Army PM offices.

5. The availability of a staff of highly skilled/experienced engineers was critically important.

Development was greatly facilitated by situations in which there had been a long history of technical work on the subject matter in question so that very skilled personnel were available. This was true, for example, at Chrysler for armored vehicles, at Hughes and at the NASA research centers for rotorcraft, at Raytheon and Lockheed Martin for missiles, at Aberdeen Proving Ground (APG) in ballistics, and at the Night Vision Laboratory (NVL)[10] for IR technology for all the systems studied. These groups were able to move quickly into new, related programs.

One of the distinguishing characteristics of government laboratories is their ability to sustain efforts over lengthy periods without frequent staff turnover, in contrast to the turnover in graduate students and post-doctoral fellows at

---

[10] NVL is now known as the U.S. Army RDECOM CERDEC Night Vision and Electronic Sensors Directorate (NVESD).

universities. This stability at the in-house laboratories developed depth of expertise. The potential downside is the possibility of laboratory staffs becoming parochial; this is a challenge that management must continually address. Also necessary is a supportive and patient management environment that champions the programs.

6. Having the right equipment and facilities is essential to enable the technical staff to carry out the work effectively and efficiently.

Development of these weapons systems required the use of some very sophisticated research equipment, often lodged in special-purpose buildings, ranges, and the like. For example, the Abrams program called for gun ranges, armor testing ranges, special facilities for testing armor and munitions containing depleted uranium (DU), test tracks, materials laboratories, and visualization techniques for measuring the behavior of munitions at very high speeds and during penetration of targets. Also, much of the work in the four systems we studied relied on advanced computers for modeling physical phenomena, such as the aeromechanics of the helicopter, finite element analysis of the composite sabot for the Abrams' kinetic energy rounds, and firing tests.

The Army did not conduct its S&T program exclusively in its own facilities: NASA and the DOE national laboratories also made important contributions. The co-located Army groups at three NASA sites took advantage of NASA expertise and special facilities. Most of the basic research for helicopters was done by Army staff at these NASA sites. The DOE laboratories were particularly helpful in developing the use of DU for the Abrams tank.

Army facilities are used not just by government staff but also by the contractors charged with producing a given weapons system. In particular, industry uses Army facilities for performance tests. It has long been the practice that new ground combat vehicles be tested at APG, for example. Also, the test facilities and laboratories at Redstone Arsenal in Alabama were used regularly by missile manufacturers.

The Army had available such special facilities and equipment for its and industry's use because substantial, long-term investments had been made and partnerships had been established. In some cases the facilities and equipment had been established and continually upgraded for decades—some dating from before World War II. Many of the facilities were and are unique. The skilled staff discussed in Finding 5 were an important factor in developing and maintaining these resources. Permanent Army installations with stable staffing have enabled this expertise to build up over time and thus to take maximum advantage of the capital investments.

7. The user communities were intimately involved in development of the weapons systems.

The contributions of Fort Knox in the Abrams program and Fort Benning in the Javelin program are clear examples of the important role the user played. The user worked with the technologists to define requirements; continuing discussions during the development phases were useful and productive. The ultimate user, the warfighter, also helped defend the programs when they were in trouble; when difficulties arise that may threaten the continuance of a needed weapons program, the full support of the user makes a very large difference.

8. The systems had key technologies in common.

Our studies found many technologies that are used in more than one weapons system. These include both technologies that aid in systems development and technologies that form actual components of the systems. M&S, for instance, was used in the development of each of the four *Hindsight Revisited* systems. M&S enables the investigator to study many different experiments on the computer thereby expanding the scope, shortening the time, and reducing the cost of development work. Infrared (IR) technology was similarly ubiquitous. Both the Abrams and the Apache made use of IR for night vision, while the Stinger and the Javelin both depended on it to seek intended targets.

Hindsight Revisited and the Original Project Hindsight

The *Hindsight Revisited* reports were inspired in part by the original *Project Hindsight*. In 1965, the Director, Defense Research & Engineering (DDR&E) Harold Brown launched *Project Hindsight*, a study of the development of 20 weapons systems from 1945–1963. *Project Hindsight* was completed in 1969 (and is referred to hereafter as DOD69).[11] The two studies have significant differences. Nevertheless, DOD69's purpose was broadly similar to that of *Hindsight Revisited*: to find out who funded the work, who carried it out, and what factors in the environment for technology development were important for success. A comparison reveals that their findings are largely in accord on the fundamental issue of what factors promote the development and successful utilization of defense S&T.

The Differences

Though DOD69 was an inspiration for *Hindsight Revisited*, there are, to be sure, significant differences. Whereas the *Hindsight Revisited* studies were conducted by the present authors through one-on-one interviews and correspondence and went deeply into the development of four Army weapons systems, the DOD69 study reviewed

---

[11] Office of the Director of Defense Research and Engineering, *Project Hindsight: Final Report* (Washington, D.C.: Department of Defense, 1969).

20 weapons systems across the Services and was conducted by eight teams, each with 5–20 participants. Further, in the 36 years between the two studies there have been some significant changes in the way technology is developed and fielded. For instance, the integrated circuit and the supercomputer, resources unavailable to technologists in the 1945–1963 timeframe, emerged during or before the development of the four systems studied in *Hindsight Revisited*. Further, the acquisition process has changed, with the roles of Program Executive Officers (PEOs) and PMs more formalized and institutionalized in the years since 1963.

Some differences are also worth noting on the two studies' findings related to basic research. In part, these stem from a difference in focus. DOD69 was driven in part by Congress, which had raised questions about the "efficiency of management and the overall payoff from the Defense Sciences part of the RDT&E [Research Development Test and Evaluation] budget" (what we refer to as the basic research or 6.1 category).[12] In the 1960s there seems to have been some question as to its importance to weapons systems development. There was also some uncertainty as to whether effective technology development came from problem-oriented research or somehow arose from less directed work. The role of basic research was not well understood and DOD management of it was in question. *Hindsight Revisited*'s brief was more limited in this respect. As discussed earlier, our studies concentrated on changes in performance brought about by technical advances as compared to the technology in the immediate predecessor systems. Most of the basic science enabling these advances was performed prior to the time

---

[12] Harold Brown, Letter to the Assistant Secretary of the Army (R&D), the Assistant Secretary of the Navy (R&D), and the Assistant Secretary of the Air Force (R&D), 6 July 1965 in *Project Hindsight: Final Report*, 135.

the systems under study began to be developed. Thus we did not, for the most part, study details of the role of that underlying science. As indicated by the CTE definition and as shown in Figure 1, our own reports are largely devoted to the later stages of the basic research-to-engineering development continuum (6.2 and beyond).

Yet when *Hindsight Revisited* did touch on basic research, its findings differed somewhat with DOD69 on one important point. DOD69 concluded that the most useful role of science (as opposed to technology) was to explain the basis of the phenomena being studied. *Hindsight Revisited* found that work done in basic research does not just provide explanations for phenomena but directly supplies the foundations for the systems of interest. We also found that the basic research has usually been done well before the launch of the weapons program under study rather than during the development period (as seen in Figure 2). Exceptions are when basic research is necessary to remove obstacles that occur during the development timeline. For example, in recent years both the Crusader howitzer's liquid propellant gun system and the electromagnetic gun returned to 6.1 for additional fundamental study.

The Similarities

As stated above, DOD69 and *Hindsight Revisited* had broadly similar objectives. DOD69 had as specific goals to:

- identify management factors that will assure DOD "research and technology programs will be productive and that program results will be utilized"; and
- measure the overall increase in cost-effectiveness in the current generation of weapons systems compared to that of their predecessors and determine what part of the effectiveness can be attributed to DOD S&T investment.

Both *Hindsight Revisited* and DOD69 focused on specific technological advances. In DOD69 these were called Research or Exploratory Development (RXD) Events; in *Hindsight Revisited* they are called CTEs. A RXD Event was limited to the actual technical work, such as the conception of an idea, the design of a new component, and the initial demonstration of military utility. The CTE was the basic unit of analysis for *Hindsight Revisited*—we sought to determine what technical advances were important and then examine the factors that led to them. Unlike RXD Events, CTEs encompassed key technical management decisions as well as some significant technical accomplishments that were not adopted for use in the weapons system or were in new versions of the system not yet fielded.

Based on their study of CTEs and RXD Events, *Hindsight Revisited* and DOD69 both found that in-house laboratories and industry made the greatest contributions to defense S&T. DOD69 found in-house laboratories and industry each contributed roughly 45 percent of the RXD Events, with academia making up the difference. *Hindsight Revisited*'s findings were similar, though we tracked joint government/industry CTEs while DOD69 did not consider joint RXD Events. As seen earlier, *Hindsight Revisited* found that government laboratories and industry together contributed about 90 percent of the CTEs. We attributed about 40 percent of the CTEs directly to in-house laboratories, about 20 percent to joint government/industry efforts, and about 30 percent to industry alone. The balance fell into the category of managerial decisions or contributions from allied nations and academia. Further, we found that development of two systems, the Abrams and the Apache, was led by in-house contributions; for the two missiles, contributions were led by industry.

Hindsight Revisited's analysis of CTEs went beyond DOD69's analysis of RXD Events in some areas of qualitative analysis. *Hindsight Revisited* included discussions of the in-house laboratory environment in terms of management support, funding, technical equipment, facilities, and the like. We looked at the value of past investments in the technical areas and the extent to which maintaining an experienced staff is important. DOD69 did not emphasize these areas.

Both studies found that the DOD funded all, or nearly all, of the S&T for the systems under review. The only exceptions were some work performed by our international partners and transferred to the Army. DOD69 found that most new technology (as opposed to science) in the DOD S&T program was a result of problem-oriented research, either in generic research for a group of like systems or to address specific challenges in the development of a specific system.[13] Today there is no question about this issue: as borne out by *Hindsight Revisited*, some basic research (6.1) and all applied research (6.2) and advanced development (6.3) in the Army is clearly driven by the weapons problems at hand.

Both studies found that the same technologies were used in multiple weapons systems. We will discuss this finding at length for *Hindsight Revisited* in Chapter III. This feature of DOD S&T frustrated one of DOD69's main goals: to compute a value-cost index for its RXD events. Sometimes single advances, like the development of the integrated circuit, were so widely applied that apportioning the expenses or benefits to any one system was impossible.

Both *Hindsight Revisited* and DOD69 found that the effective utilization of new technology in a weapons system depends on a close relationship between the user and those

---

[13] It appears that DOD69's focus on whether research is problem-oriented or from random, curiosity-driven research arose in the 1960s from external critics of the DOD program.

doing the technical work, both in government laboratories and in industry. For example, the *Hindsight Revisited* studies showed the important roles played by the Armor Center/School at Fort Knox, KY, in the development of the Abrams tank and the Infantry Center/School at Fort Benning, GA, in the development of the Javelin antitank missile. These representatives of the warfighter stated their needs and then provided strong advocacy for the programs, particularly when the programs encountered challenges.

Both *Hindsight Revisited* and DOD69 found that a very important technology transfer mechanism was informal person-to-person contacts. *Hindsight Revisited* found that this is even truer today, as technology has made communication faster and more efficient. The internet for e-mail and file transfer as well as specialized networks of scientific computers enable teams of geographically dispersed workers to carry out distributed collaborative research. Indeed, *Hindsight Revisited* placed more emphasis than did DOD69 on the general importance of teaming among the participants to develop technology as well as integrate the many technologies of all the components of a system. It is important to note that the introduction of the PEO/PM since 1969 had a large impact in this area and helped facilitate such additional technology collaboration and transfer measures as moving personnel with the project from laboratory to laboratory or from laboratory to industry and creating integrated project and process teams. It should also be noted that today there is increased emphasis on formal technology transfer as evidenced by Technology Transfer Agreements with the PM office. Here, there is a signed agreement that details products to be delivered and schedules and metrics to demonstrate maturity.

Beyond the *Hindsight* Studies

Project Hindsight and *Hindsight Revisited* have one final commonality: they are not idle exercises. As discussed in the introduction to this volume, such retrospective studies have a compelling logic—examining and understanding past successes can guide the successes of the future. What guidance, then, do we take from the *Hindsight Revisited* findings summarized above? Without diminishing the importance of the others, we have selected two of these findings to expand upon in the coming chapters. First, though the Army fields many diverse, complex weapons systems, certain groups of technologies apply to many or all of them. We saw such common technology elements spring up in each of the *Hindsight Revisited* systems. These common technologies are clear areas in which the Army S&T program should maintain excellence and in which it could hope to see its investment pay off throughout the force. In Chapter III, we will discuss some such technologies that appeared in the *Hindsight Revisited* reports and that will continue to be important to the Army. Second, *Hindsight Revisited* showed us the vital importance of a robust Army laboratory system. Chapter IV addresses how the Army can ensure that this essential factor for past success is available to support the successes the Service will demand in the years to come.

Chapter 3

Technology Areas of Broad Impact

Though the *Hindsight Revisited* effort detailed in Chapter II covered four disparate weapons systems, some important technologies were common to all. This chapter will take up several of the most significant of these common elements and address their importance for the development of future weapons systems.

For convenience we divide these technologies into two groups: those that enable the development of weapons systems and those that are contained in the products of the development process; i.e., weapons systems and warfighting equipment and their subsystems and components. In the former group, three enabling technical factors stood out from the *Hindsight Revisited* studies as especially important to future Army S&T efforts: the use of M&S, the use of high performance computers, and the use of human-systems integration. In the latter group, we will touch briefly on three technical areas—gun technologies, infrared sensors, and power sources—that bore fruit for more than one of the *Hindsight Revisited* systems and will continue to reward Army S&T investment with payoffs for many future platforms. We will then address one particular area of great future importance for the Army: those technologies bearing on network-centric operations.

Enabling Technologies

The Army S&T program has developed or adapted many processes and capabilities that were and are essential to the development of new military systems. Among them are M&S and high performance computing, which now play a major role in laboratory operations, and human-systems integration, which is made vital by the increasing demands placed on soldiers by modern weapons platforms and the ever-greater flood of information generated by state-of-the-art sensors and communications networks. The *Hindsight Revisited* systems all benefited in varying degrees from these technologies, and all three of these are areas in which the Army will continue to reap good returns on its S&T investment.

Modeling and Simulation

The impact of M&S is difficult to quantify on the battlefield, but it is an important S&T capability, one that played a key role in the development of the Abrams, the Apache, the Stinger, and the Javelin. Use of this technology enhances the effectiveness of the Army S&T program, allowing researchers to efficiently investigate complex problems. M&S will become even more important in the future, as the development of ever-more complex systems calls for a detailed understanding of physical phenomena and of the prospective performance of new systems on the battlefield.

The terms *modeling* and *simulation* deserve some explanation. Though often used together, there is a distinct difference between the two. As defined in the DOD M&S Glossary, modeling is the "application of a standard, rigorous, structured methodology to create and validate a physical, mathematical, or otherwise logical representation of a system, entity, phenomenon, or process."[14] Models of both physical systems (e.g., the effect of a given density of armor on a penetrating round) and non-physical systems (e.g., the effect of a given volume of information on the speed of human decision-making) are possible. Models of systems of systems may be composed of numerous physical and non-physical relationships.

In the same M&S Glossary, a simulation is defined as "a method for implementing a model over time." Computer simulations are based on models or mathematical formulae that describe a real system or phenomenon.[15] When executed, the computer program, or simulation, shows how the system works and, by changing variables, it is possible to analyze what effect changes might have on the real system.

> Electronic displays are critical components of simulations, in that they provide a medium for viewing representations of simulations as well as data arising from computations, sensors, telecommunications links, and so on. Displays are themselves a technology area with application across the force. Displays have proliferated on the battlefield—on computer monitors, instrument panels, on heads-up displays on helmets, and many other devices—and will only become more ubiquitous. Thus they can be exploited to integrate simulations throughout the battlefield.

---

[14] *DoD Modeling and Simulation (M&S) Glossary*, DOD 5000.59-M, Undersecretary of Defense for Acquisition Technology, January 1998. Accessed 7 January 2007 at <http://www.dtic mil/whs/directives/corres/pdf/500059m_0198/p500059m.pdf>.

[15] *NOVA Science in the News Glossary*, Australian Academy of Sciences. Accessed 7 January 2007 at <http://www.science.org.au/nova/glossary.htm>.

Simulation in the military falls into three categories: live, in which real people use live equipment (fielded, prototype, or surrogate) in the real world; virtual, where real people use mock-ups of equipment (a simulator[16]) in a simulated world; and constructive, where everything is simulated, such that simulated people operate simulated weapons systems.[17] In every case underlying models relate cause and effect by "if-then" statements. An example is: "if a soldier is illuminated by a laser spot representing a rifle shot, then he is a casualty." Virtual simulations inject human motor control skills (e.g., driving a tank), decision skills (e.g., committing a unit into action), or communication skills (e.g., live commander communicating with live, virtual, and constructive subordinates) into the simulation. Force-on-force computer models that simulate a battlefield with friendly and opposing forces are examples of constructive simulations. Real people stimulate (make inputs to) constructive simulations, but are not involved in determining the outcomes. Often, these categories have no clear separation between them. Additionally, advances in today's technologies now call for an additional category for simulated people to operate real equipment (e.g., simulated commanders having control of units that include live smart robotic vehicles).

Each of the *Hindsight Revisited* systems made use of M&S, especially physical modeling. An example from our study of the Abrams tank is the M&S of live fire testing. In the models, the ballistics of the flight of the incoming round, of the penetration of the armor, the spraying of fragments in the interior, and the impact of the fragments

---

[16] A simulator is a device that duplicates the essential features of a task situation and provides for direct human operation, such as a mock-up of an Abrams tank with crew stations, vision blocks projecting simulated scenes, speakers projecting vehicle and battlefield sounds, etc.

[17] *DoD Modeling and Simulation (M&S) Glossary.*

on the crew and components are all represented by mathematical models of the physics. Other variables include the angle of attack and the nature of the round (high-explosive or kinetic energy). Since many of the variables are not single-valued but exhibit a natural variation, probabilistic models are used to provide probability distributions for these variables. The product is a distribution of results—the probability of penetration, the probability of hitting an item in the interior, the probability of disabling the item hit and the probability of the tank being put out of action. These models are used to design firing tests and predict their outcomes. By use of the models, many tests can be simulated rapidly on the computer so that the actual live fire tests can be focused on the most significant of the many simulated scenarios. Fewer real tests are required thereby speeding up development and reducing cost.

While physical, laboratory-focused M&S has been and will continue to be vital to Army S&T efforts, the remaining part of this section will focus largely on M&S applications that encompass not just weapons development but also training and battlefield operations. M&S makes it possible for researchers to examine, at a comparatively early stage in the development cycle for comparatively little money, many of the growing number of variables that can impact the performance of a weapons system or even a battlefield operational concept. The increasing complexity of today's weapons systems and of the modern battlefield puts an especially high premium on such technologies.

The Army has a long history of using M&S in technology development, analytical efforts, acquisition, training, and testing. The Army's first simulator, the Link Flight Trainer, was developed in the 1930s and used for flight training during WWII. In the 1950s, NASA and the U.S. military began developing complex simulators for training. The Apollo 13 simulator made headlines for its use in saving the Apollo 13 crew in 1970. By the end of the 1970s, the use of simulators for training was widespread, and the use of simulations for analysis and technology development was beginning to take hold.

Early simulators were stand-alone systems designed for single task training purposes such as driving a tank. In 1988, DARPA initiated a program called Simulator Networking (SIMNET) that would join multiple tank simulators over a network such that each could see each other and as a team of tanks they could detect, engage, and destroy enemy targets.[18] This enabled commanders to focus on unit training in addition to individual crew training.

Prior to SIMNET, computer-generated forces demonstrated mostly predictable, robotic behavior. Now, in addition to these constructive forces, a limited number of friendly and opposing forces alike could reflect the more accurate, variable behavior of the live crews manning their simulators. With the added use of combat gear and realistic simulator interiors, crews increasingly believed the training was "real" and not "virtual." This increased realism was cited in Operation Desert Storm after-action reports as contributing to U.S. military success.[19] These findings prompted the Army to continue to develop SIMNET-enabled training centers as a means for preparing for war. Building on this success, the Army later developed and replaced SIMNET systems with an enhanced networked simulation-enabled training system called the Combined

---

[18]"PC-Based Technology Invades Simulation," U.S. Army Program Executive Office for Simulation, Training, and Instrumentation. Accessed 5 November 2006 at <http://www.peostri.army mil/ PRODUCTS/PC_BASED_TECH/ >.
[19] "Operation Desert Storm, War Offers Important Insights Into Army and Marine Corps Training Needs," GAO Report # GAO/NSIAD-92-240, August 1992. Accessed 5 November 2006 at <http://archive.gao.gov/d35t11/147541.pdf>.

Arms Tactical Trainer.[20] This system continues in use today.

SIMNET was the predecessor of the Distributed Interactive Simulation (DIS) protocols that linked a wider variety of geographically distributed simulations and weapons system simulators such as trucks, helicopters, fighters, ships, and soldiers to create single virtual battlespaces with entity interactions occurring in real time. For almost a decade, distributed simulations would remain focused on linking constructive models and simulations and virtual simulators The rapid development of M&S accelerated its use beyond training. For example, in the summer of 1990, a computer-based war game called Operation Internal Look was used by General Norman Schwarzkopf and his staff at the U.S. Central Command to run through scenarios of potential conflict in Iraq. Immediately after the invasion of Kuwait, the function of Internal Look changed from planning to execution; it was used to run variations of the real combat scenario.[21]

In the 1990s, simulation became a common tool for military technology development, analytical efforts, acquisition, training, and testing. Army laboratories linked their engineering models, or at least close representations of their conceptual warfighting systems, to force-on-force simulations. This enabled the technical staff to see how their new systems would affect the battle space. The use of distributed simulators also allowed them to conduct experiments with humans-in-the-loop to evaluate individual and unit system performance.

For almost a decade after the inception of SIMNET, distributed simulations remained focused on linking

---

[20] Further information on CATTs is available at
<http://www.peostri.army.mil/PM-CATT/home.jsp>.
[21] Kevin Kelly, *Out of Control* (New York: Perseus Books Group, 1995). Accessed 5 November 2006 at <http://www kk.org/outofcontrol/ch22-e html >.

constructive models and simulations and virtual simulators. DIS protocols could also link live simulations, but that meant that the constructive and virtual simulations would have to operate in real time—to keep up with the participating live players—and live players would have to be instrumented to be "emulated" within the simulation. Prior to 2000, shortcomings in processing power, bandwidth rates, and instrumentation for live players kept live simulations separate from integrated virtual-constructive simulations. In the past six years, by exploiting advances in high performance computing (HPC) and infrastructures such as the Defense Research and Engineering Network (DREN), technological advances have made possible a more transparent integration of live forces with models, simulations, and simulators over widely distributed networks. The DREN supports the linkage of disparate laboratory, training, experimentation, and testing facilities, enabling highly complex Service and joint live-virtual-constructive (LVC) M&S mission environments.[22] The environments can present an operational battlespace that is as close to war as possible, with many of the complexities of joint operations, without being inhibited by safety, health, and environmental constraints. This is invaluable for accurately assessing technologies and evaluating tactics, techniques, and procedures.

Among the things such advanced simulation technologies make possible are constructive simulations of communications systems. Systems at Fort Monmouth can be linked to FCS simulators at Fort Knox, live dismounted forces at Fort Benning, a live command and control element at Fort Leavenworth, an Air Force Tactical Control

---

[22] Mike Cast, "Army Tests Move to Virtual Proving Ground," *National Defense*, November 2001. Accessed 5 November 2006 at <http://www.nationaldefensemagazine.org/ issues/2001/Nov/Army_Tests htm>.

Party at Eglin Air Force Base, and an AC-130 simulator at Hurlburt Field. With this setup, joint close air support and Army tactical communications systems can be assessed, among other joint and Army tasks and capabilities. Although setting up such distributed simulations is time- and resource- intensive, it is still more cost effective than trying to conduct the actual live exercises at a single range facility. Looking towards the future, such capabilities enable assessments of large FCS forces in joint environments well before these forces are in being.

Another recent advance in M&S that is not time and resource intensive is agent-based simulation (ABS). In conventional computer simulations, entities (like simulated enemy tanks) follow finite rules programmed in by the user. Agent-based simulations differ in that agents, or intelligent computational entities, can alter their behavior within the simulation based on information provided by the simulator.[23] The range of an agent's behavior varies during the simulation, with its end points being defined by simple "if-then" case-based reasoning and very complex machine learning algorithms (e.g., genetic algorithms).

Agent-based simulations can better emulate a very complex, dynamic world of learning entities; conventional simulations have entities that behave only as the modeler has programmed. For example, the U.S. Marine Corps uses the ABS Map Aware Non-uniform Automata (MANA) in an effort entitled Project Albert. MANA is specifically designed to model combat and how soldiers interact and has the ability to model information systems that provide agents with varying amounts of information and situational awareness.[24] These models then produce the desired

---

[23] *Agent Rescue Emergency Simulator Glossary.* Accessed 7 January 2007 at <http://pages.cpsc.ucalgary.ca/~kidney/ARES/7_Glossary/ 7_Glossary html>.
[24] MAJ Patrick M. Downes, LTC Michael J. Kwinn Jr., and Donald E. Brown, "Using Agent-Based Modeling and Human-In-The-Loop

simulation. Three significant benefits are (1) ease in setting up and operating, (2) the semi-automated agents themselves, and (3) the ability to model situational awareness. System performance parameters can be changed to represent future capabilities of developing technologies. Setting up and generating usable output from an ABS may take weeks; it would take months to do the same for a typical Army simulations of this kind.[25] ABS is now being used by Training and Doctrine Command (TRADOC) in their Analysis Centers and Army Battle Laboratories.

The Army also has moved to tap advances in the commercial world, particularly the entertainment industry's increasingly realistic computer games and movie special effects. In the past few years, commercial games[26] have increased tremendously in fidelity and in the ability to immerse players in the game environment. Although games lack physics-based reality they nonetheless follow on a set of rules based on "if-then" sequences. The Army has found value in using games for military education (as opposed to military training). Recent advances in game fidelity have begun to extend their use in experiment and test planning. Among the initiatives the Army uses to exploit the

---

Simulation to Analyze Army Acquisition Programs," *Proceedings of the 2004 Winter Simulation Conference*, INFORMS Simulation Society. Accessed 5 November 2006 at <http://www.informs-cs.org/wsc04papers/127.pdf>.

[25] LTC Thomas Cioppa, briefing slides, Training and Doctrine Command Analysis Center, Monterey, CA, 14 January 2005.

[26] The relationship between computer games and simulations is as follows: "Whereas a simulation is a serious attempt to accurately represent a real phenomenon in another, more malleable form, a game is an artistically simplified representation of a phenomenon. The simulation designer simplifies reluctantly and only as a concession to material and intellectual limitations. The game designer simplifies deliberately in order to focus the player's attention on those factors the designer judges to be important." Chris Crawford, *The Art of Computer Game Design*, online book accessed 5 November 2006 at <http://vancouver.wsu.edu/fac/peabody/game-book/Chapter 1.html>.

expertise of the commercial gaming community is the Institute for Creative Technologies at the University of Southern California, a 6.1-supported Center of Excellence for virtual reality and computer simulation research.

Despite these decades of steady advances in M&S, there is room for improvement. Current computational systems and bandwidth rates continue to limit the number of player entities and command, control, communication, computer, intelligence, surveillance, and reconnaissance (C4ISR)[27] nodes that can appear in a distributed LVC simulation in real time. Computational power also limits the real-time integration of highly complex models into a large force-on-force simulation using live players. One example is the integration of a very complex engineering model of shot-line analysis to evaluate tank damage for every engagement in the distributed LVC simulation. As computing power and bandwidth rates grow, so will the ability of the Army laboratories to link their engineering models of advanced technologies and conceptual warfighting systems into the distributed simulation environment of the Army training, analysis, and acquisition communities.

---

[27] Command, Control, Communications, and Computer systems are integrated systems of doctrine, procedures, organizational structures, personnel, equipment, facilities, and communications designed to support a commander's exercise of command and control across the range of military operations. Intelligence, Surveillance, and Reconnaissance are activities that synchronize and integrate the planning and operation of sensors, assets, and processing, exploitation, and dissemination systems in direct support of current and future operations. These are integrated as intelligence and operations functions. *Department of Defense Dictionary of Military and Associated Terms*, Joint Publication 1-02, 12 April 2001 (as amended through 9 November 2006). Accessed 7 January 2006 at <http://www.dtic mil/doctrine/jel/new_pubs/jp1_02.pdf>.

High Performance Computing

High performance computing was important to the development of each of the *Hindsight Revisited* systems, not least because it enabled some of the M&S described above. As weapons systems become more complex, HPC will only grow in importance as an enabling technology in weapons development. Researchers use HPC capabilities for many purposes. Such computing power makes technical work quicker and much more productive and enables experiments that cannot be done in the laboratory. Very large computer programs for computational fluid dynamics, materials by design, and structural mechanics, for example, are now replacing much of the experimental work done in the past, saving time and money. The Army S&T program has and will continue to require HPC to function effectively; investment in this tool is likely to aid the development of many sorts of weapons systems in the future.

HPC makes use of special hardware (these extremely fast machines are referred to as supercomputers) and software, plus specialists in writing algorithms and software for machines that can handle very large amounts of data and a great number of sequential or parallel computations. Only a few kinds of demanding problems are candidates to use such capabilities. Such endeavors are highly specialized. The growth in high performance computing was fostered by large investments, mostly from the Federal government. The principal thrust initially came from the nuclear weapons program, which needed HPC to model the effects of nuclear explosions. Subsequently, disciplines such as meteorology needed to process vast amounts of data from space, atmospheric and ground stations to develop forecasts. The number of high performance computers is relatively small; they are usually in centers specializing in the work. Note, though, that HPC capabilities limited to just a few research centers in one generation may be available in basic consumer computers a few generations later. For example, the first programmable electronic numerical integrator and calculator (ENIAC) was regarded as a super calculator by contemporaries then using mechanical devices or doing calculations by hand, but its capabilities are trivial compared to today's machines.

HPC was especially important to enabling M&S for the *Hindsight Revisited* systems. Such computing resources were needed, for instance, to analyze the millions of variables involved in a model of a projectile striking the Abrams' crew compartment, or of the dynamic interactions between the Apache's rotor wash and its horizontal stabilator. The Javelin benefited from the High Performance Computing Modernization Program (HPCMP), which was initiated in 1992 in response to congressional direction to modernize the DOD laboratories' HPC capabilities. The thrust of the HPCMP is to provide DOD scientists and engineers with state-of-the-art computing resources and training. Many military systems have benefited from this effort. For the Javelin, HPCMP enabled advanced design concepts, improved and faster modification programs, higher fidelity simulations, and more efficient tests.

The military's largest computer programs are now run in supercomputer centers such as the Defense High Performance Computing Major Shared Resource Centers (MSRCs). Four MSRCs in DOD currently operate large HPC systems that are available to the entire DOD HPC community. Each MSRC provides a complete HPC environment, including hardware, software, data storage, and computational expertise. In addition to the MSRCs, more modest-sized local systems are in place where there is a significant need for HPC capability. For analysis efforts that require extremely intense computing power, high performance computers at the centers are linked together using the DREN or the Secure DREN.

With its MSRC, the Army Research Laboratory (ARL) can now interactively visualize enormous data sets from calculations consisting of billions of computational cells. Previously, calculations of this magnitude would have to be run in the batch mode, sometimes for days. Using the supercomputer an investigator can not only run the calculations in a short time, but he or she can also visualize the results in real time. In this way the investigator can decide how to pose the next step in the problem without a long wait. It is no small task to convert the very large data sets of the results into pictures on the screen. Special programs are needed for this as well.

The Army is using and will continue to use such centers to meet future weapons development challenges. One example is the recent analysis of armor options for Abrams side armor and High-Mobility Multipurpose Wheeled Vehicle (HMMWV) appliqué add-on armor, developments needed in part to respond to threats in Iraq.[28] This work is similar to the M&S discussed in the Abrams *Hindsight Revisited* report, but significantly enhanced computing power and bandwidth rates provide an opportunity for developing even more complex models and distributed simulations. To simulate the details of live-fire tests on the Abrams or other platforms, the Army at APG has built models containing as much of the physics and geometry as possible for various angles of attack and points of impact. In reality, most of the variables are not represented by specific discrete values but rather exhibit a probability distribution of possible values. There will be variations— slight but significant—in the velocity, shape, and mass of the incoming rounds. Penetration of the vehicle will vary depending on variations in the manufacture of the armor. Assuming penetration of the armor, the spatial distribution,

---

[28] David Kleponis, "Armor Research Provides Safer Tactical Vehicles," *ARL MSRC Link*, Fall 2005, 20.

number, masses, and velocities of the behind-armor fragments will vary. These uncertainties translate into a distribution of possible results rather than a single result. The HPC-run calculations will produce probabilities for hitting targets within the vehicle and probabilities as to whether the hit results in damage, minor or disabling, and whether the vehicle remains in operation or is put out of action. The simulation is checked by running a few live firings. Then the model is used both in designing live-fire tests, predicting results, and changing design features of the vehicle. In the early days of this modeling, as in the early modeling of the Abrams tank, only a few variables could be modeled and these without a probability distribution of input and output variables; with more powerful computers more detail and greater accuracy can be produced.

Such HPC activities benefit systems in the field today. Future weapons systems also will benefit from HPC. High-end computers can model the use of high power microwaves for non-lethal attack of electronics, like those used to trigger improvised explosive devices (IEDs).[29] Others uses include systems-of-systems simulation, basic and applied research in plasma physics, turbulence modeling, molecular engineering, and modeling the design and reactions of armors and high-energy materials.[30] FCS in particular will benefit from HPC capabilities. One area in particular in which it will do so is the M&S of its C4ISR system. The battlefield network for FCS will include tens of thousands of nodes. For networks of this size, it is a daunting task to analyze message traffic in terms of throughput, bandwidth rates, impact of loss of nodes, loss of packets, and other criteria. It is extremely inefficient and

---

[29] Keith Cartwright, "Directed Energy Weapons Subject of Challenge Project," *ARL MSRC Link*, Fall 2005, 21.
[30] "Components of the HPCMP," Department of Defense High Performance Computing Modernization Program. Accessed 5 November 2006 at <http://www.hpcmo hpc mil/>.

time consuming to conduct detailed analyses of these large scale systems of systems on desktop computers, so it is imperative to use HPC assets.

The MSRCs ensure that Army scientists and engineers across the country have access to the best HPC and expertise available. Utilization of these assets cuts defense system cost by reducing the design cycle and dependence on expensive live experiments and tests. The Army should continue to invest in and leverage HPCMP resources and capabilities.

Human Systems Integration

In all our *Hindsight Revisited* studies we found that the interface between the soldier and the weapons system had to be carefully engineered. The placement, color contrast, and symbology of displays; placement of controls and the ease with which they are manipulated; methods and levels of training demanded by a system—these and other similar factors are very important to developing a successful weapons system. Issues of this sort are dealt with through the discipline of Human Systems Integration (HSI), also referred to as Warfighter-Machine Interface (WMI). HSI means systematically factoring human considerations into the design, acquisition, and support of systems. HSI has been important to the Army S&T program in the past, and with the introduction of more and more complex technology, will continue to be in the future.

The Army has devoted increasing attention to HSI.[31] In the 1960s, 70s, and early 80s, the Army introduced into the force numerous advanced technology weapons systems, designed to increase Army combat power and readiness. During this modernization, the Army identified two persistent problems: fielded systems did not perform as

[31] "MANPRINT History," U.S. Army MANPRINT Directorate. Accessed 5 November 2006 at <http://www.manprint.army.mil/manprint/mp-history.asp>.

predicted (for example, probabilities of hit were half of those seen in laboratory tests) because of poor human-system design and the new system often required more, and more highly skilled, soldiers (operators, maintainers, and support personnel) than the system being replaced. The Army recognized it could not continue to accept poor system performance nor could it recruit sufficient numbers of highly skilled personnel or implement longer, costlier training programs. In 1980, prompted by high-level concern within the Army, a study was done on the effects of increases in weapons complexity on mobilization, readiness, and sustainability.[32] The study concluded that human performance assessments were often not made and, if made, were too late to influence the design of the system.

In 1982 the Army leadership directed that a manpower and personnel integration program be initiated. The term "MANPRINT" (MANpower and PeRsonnel INTegration) came to be used to designate the program. The Army's MANPRINT program considers both soldier and unit needs throughout the system acquisition process and life cycle. MANPRINT is the Army's program for improving the effectiveness of system performance at minimum costs for personnel, maintenance, and repairs. This design objective considers seven key areas: manpower, personnel, training, human factors engineering, safety, health hazards, and soldier survivability. MANPRINT optimizes total system performance and minimizes cost. Throughout the design and development phases, MANPRINT ensures that:

- system operation, maintenance, training, and support requirements are matched to personnel availability;
- systems become increasingly soldier-centered, -trainable, -reliable, and -maintainable;

---

[32] Ibid.

- life cycle costs are reduced through minimizing or eliminating specialized skills and tools for user-level maintenance; and
- total system performance is optimized at minimal life cycle costs by proper assignment of functions to man or machine.[33]

The *Hindsight Revisited* studies touched on several important examples of the application of HSI. One such was the difference in classification between the Redeye and Stinger anti-aircraft missile systems. The Redeye was fielded as an "all-arms" weapon, meaning that any soldier could fire it with little training. For the Stinger on the other hand, after consideration of the weapon's capabilities and the training needed to operate it effectively, the Army determined that the missile required gunners who were trained Air Defense personnel. Another example of HSI contributed significantly to the battlefield effectiveness of the Apache. Both crew members' helmets are fitted with a monocle-like heads-up display that gives piloting and target information. Using flight simulators, the symbols that appear on this display were carefully designed with regard to what information was essential and how that information should be presented to communicate it as quickly and with as little distraction as possible.

HSI and the MANPRINT program will continue to play an important role in the development of future systems. Army technologies and systems will continue to become more complex and warfighters will have to cope with increasing amounts of technology in the field. To cite just one broad area as an example, HSI will have an especially key role in enabling information-intensive operations. As advanced communications devices and situational awareness technologies proliferate, the individual soldier is

---

[33] Ibid.

expected to provide and react to an ever greater array of information. The more that is asked of the soldier, the greater the benefit of a development process that early on factors in the soldier's abilities and needs.

Technologies for Fielded Systems

M&S, HPC, and HSI all enable the effective development of new Army capabilities. Along with these common technical enabling elements, *Hindsight Revisited* also revealed common areas of technology in the fielded systems. We focus on just three: gun technology, night vision/infrared technologies, and power sources. These technologies have contributed, and will continue to contribute, to the success of many weapons systems, and so make excellent targets for future Army S&T investment.

Gun Technology

That gun technology is important to many systems is hardly surprising: the Army is, always has been, and for the foreseeable future will continue to be, reliant on guns to execute its mission. In the *Hindsight Revisited* reports, gun technology advances gave the Abrams and the Apache important new capabilities. It is important, though, that gun technology not be taken for granted. The Army has long experience with the technologies associated with gun development, foremost among them ballistics and materials technology. The Army's long experience and evident success with gun technology, however, is not reason to assume that the work is done. Developing a more powerful gun or a more lethal projectile (or stronger, lighter armor) will continue to involve painstaking ballistics and materials research. Here, we discuss how investment in just one subset of gun technology—using interior ballistics and related materials technologies to address barrel wear and durability—could have a payoff for the development of future systems.

In order to prevent catastrophic failure and increase gun tube life, barrels must be designed with a keen understanding of the interior ballistics involved. The act of firing any gun is taxing on the barrel and barrel wear is a particularly challenging problem. For example, during the few milliseconds it takes a tank cannon to fire, the bore of the gun will experience a pressure pulse of about 100,000 pounds per square inch and a thermal pulse in excess of 2550°F and will be exposed to corrosive chemicals in the propellant.[34] During the Abrams' development, technologists at Watervliet Arsenal developed test procedures, manufacturing methods, and special electroplating techniques to improve the 105mm (and later 120mm) main gun and extend its service life. Despite the Army's efforts, the 120mm gun still fails after between 180 and 375 rounds due to barrel wear, meaning that it has experienced actual firing conditions for only a few seconds of its life.[35] This number of firings is only a fraction of the fatigue life of the barrel, adding significantly to the life cycle cost of the weapons system. Barrel wear is also a problem for smaller caliber weapons. The barrel of the M240, the primary medium machine gun of the Army and Marine Corps, can burn out after two minutes of rapid fire. This limits the combat effectiveness of the weapon during heavy engagements and burdens the machine gun team with at least one spare barrel.

Future weapons programs also face barrel wear issues. For example, the FCS program plans propose a Non-Line of Sight Cannon (NLOS-C) platform to provide indirect fire support. The NLOS-C has an objective sustained rate of fire with its 155mm cannon of 10 rounds per minute,

---

[34] Gregory N. Vigilante and Christopher P. Mulligan, "Cylindrical Magnetron Sputtering (CMS) of Coatings for Wear Life Extension in Large Caliber Cannons," *Materials and Manufacturing Processes*, eds. William S. de Rosset and Michael Audino, June 2006, 621.
[35] Ibid.

with the capability of firing four to six rounds so quickly and at such trajectories that they arrive on target at roughly the same time.[36] The 155mm cannon on the Army's current self-propelled howitzer, the M109A6 Paladin, has a sustained rate of fire of four rounds per minute.[37] This means the Army has a clear requirement for substantially more durable gun barrels. NLOS-C and other future weapons also will take advantage of advances in propellant technology that generate greater velocity and range more efficiently. These propellants, though, create higher temperatures and pressures and may also entail harsher chemical environments. Such factors can increase erosion of the internal surface of the gun. Increased firing rates and harsher internal ballistics environments demand improved materials together with innovative processing concepts. Two candidate solutions for application to future gun systems are ceramic and refractory metal gun barrel liners.

Ceramic materials offer the promise of excellent resistance to elevated temperature environments at higher hardness and lower density than steel. The idea of using ceramics is not new. A great deal of effort has been expended trying to determine the best ceramic for lining gun tubes. Silicon carbide and silicon aluminum oxynitride materials have been tried in single-shot testing and showed little wear or erosion.[38] However, when tested in a burst-fire mode, the ceramic liners cracked, so continued S&T work is needed.[39]

---

[36] "NLOS-C Demonstrator Facts," BAE Systems. Accessed 5 November 2006 at <http://www.uniteddefense.com/prod/nlos_cannon_faqs.htm>.

[37] "Paladin," U.S. Army Fact File. Accessed 5 November 2006 at <http://www.army.mil/fact_files_site/paladin/index html>.

[38] William S. de Rosset and Michael J. Audino, "Advanced Gun Barrel Materials and Manufacturing Technology Symposium—Overview and Perspective." *Materials and Manufacturing Processes*, 571.

[39] Additional materials development programs aimed at finding solutions to the cracking problem are ongoing. An interesting approach

Refractory metals are also possible gun liner materials. This is a class of metals, including tantalum, tantalum alloys, tungsten and rhenium, that is extraordinarily resistant to heat and wear. The challenge with these refracting metals and alloys, which have very high melting points, is how to apply them as liners at temperatures that will produce the required metallurgic bond but will not adversely affect the beneficial residual compressive stresses in the gun tube. Some novel processing technologies to address this issue are under evaluation.[40] High density infrared heating, magnetron sputtering, and explosive bonding are all being explored as ways to attach a refractory metal liner to a steel gun tube.

The ubiquity of guns in the Army means that gun technology needs close attention for the foreseeable future. Advances in the ballistics and materials technologies are needed to improve barrel wear characteristics to realize potential gun capability improvements. S&T investment in these and other gun technology areas will have application across the force.

Night Vision and Infrared Technologies

Sensors play a vital role in nearly everything the U.S. Army does, from navigation to combat. They are prevalent both as functional components of other systems and as stand-alone systems. Imaging and non-imaging sensors use an array of sensing approaches, including optical camera, radar, hyperspectral, acoustic, magnetic, infrared, laser, and many others. Of these disciplines, infrared technology perhaps most clearly illustrates the enduring multi-system

---

under current evaluation involves a ceramic matrix composite liner functionally graded to a metal matrix composite jacket.

[40] John D.K. Rivard, Craig A. Blue, David C. Harper, Jacob J. Stiglich, Gautham Ramachandran, and Victor K. Champagne, Jr., "High-Density Infrared Cladding of Ta on Steel." *Materials and Manufacturing Processes*, 612-617.

importance of sensor technology. Infrared technology, as shown in part by the vital role it played in the development of all four of the *Hindsight Revisited* systems, is and will continue to be of great importance to the Army S&T program.

The U.S. Army has benefited greatly from long, substantial investment in infrared technologies. The proliferation of night vision capability has led the Army to claim it "owns the night," and has given U.S. forces an important combat advantage. Speaking of the 1990-91 Persian Gulf conflict one U.S. Army general remarked "Our night vision capability provided the single greatest mismatch of the war."[41] This capability has altered, rather than merely improved, the way the Army operates. Such an overwhelming competitive advantage dictates that in many circumstances the Army would *rather* operate at night.

Some night vision devices, including most of those used by individual soldiers, are image intensifiers: they amplify ambient light to improve vision in the dark. Other devices are thermal imagers: they resolve infrared signatures into useable images and targeting information.

It is this latter group of technologies that was key to the development of the Abrams, the Apache, the Stinger, and the Javelin. The Abrams and the Apache both used Forward Looking Infrared (FLIR) sensors for navigation and targeting. These devices generated thermal images that enable nighttime operations as well as operations in smoke, fog, or other visually challenging circumstances. The Abrams has thermal viewers for the driver, gunner, and tank commander. The Apache has thermal viewers built into the TADS/PNVS sensor suite. To name just one point of overlap in the development of the sensors used by the Abrams and the Apache, both benefited from the Army's

---

[41] "NVESD About Us," Night Vision and Electronic Sensors Directorate. Accessed 5 November 2006 at <http://www.nvl.army mil/about/index.php>.

Common Module approach to fielding FLIR systems. Based on work done at NVL, the Army used FLIR Common Modules to standardize parts for use in the FLIR devices on many systems and obviate the need for developing platform-specific night vision systems. This resulted in substantial cost savings.

Infrared components are vital to the Stinger and the Javelin as well. Both the Stinger and the Javelin rely on IR seekers to find their targets. The Stinger seeker locks on to the heat signature of the engine exhaust of airplanes. Whereas the Stinger's seeker only distinguishes the target as a thermal spot, the Javelin's seeker, much like the IR systems on the Abrams and Apache, resolves the thermal signature into an image. The Javelin's command launch unit (CLU) provides a high degree of resolution; the operator looks through the CLU to locate the target, and must be able to distinguish clearly between enemies and friends. The Javelin missile round's IR system also generates an image, but only of sufficient resolution to allow the guidance system to continue to pick the target out from the small area at which the gunner aimed.

IR technologies will continue to be important to a wide variety of weapons systems. Virtually every U.S. Army combat vehicle and aircraft uses IR sensors for night operations. Such systems are widely used by individual soldiers and in soldier-portable systems like the Stinger and the Javelin. As long as the Army intends to "own the night," S&T investments in IR technologies will pay off across the force.

Power Sources

Sources of power are needed to supply electricity for the many electronic devices on the battlefield. While power supplies for platforms and fixed installations are and will continue to be important areas for Army S&T investment, power supplies for the soldier, principally batteries, also have high-leverage potential. The Army is preparing itself for the kinds of light, often dismounted, operations demanded by the kind of irregular warfare found in Iraq and Afghanistan, and it is also moving towards a future force in which a robust command and control (C2) network will extend down to the squad and soldier level. As it does so, man-portable technologies, like radios, night vision goggles, laser designators, and global positioning system (GPS) receivers will continue to proliferate. With these technologies, the weight, cost, and efficiency of batteries will be of constant and increasing concern to the Army.

When it comes to equipment for the dismounted soldier, size and weight restrictions have always been paramount concerns. The *Hindsight Revisited* studies touched on this issue in the development of the Stinger and the Javelin: the man-portable systems had to adhere to strict weight requirements. This systems engineering challenge drove cost and performance trade-offs. In the case of the Javelin, the initial weight limit had to be increased to make the weapon viable.

Today, soldiers already carry well above the target weight in equipment. The typical soldier load in Afghanistan is 92–105 pounds, while the target for the next generation soldier is 45–50 pounds.[42] Along with all their other gear, soldiers must carry enough power to last for the duration of the mission—72 hours worth is considered a combat load.[43] Reducing the weight of each individual battery unit is the most obvious way to lighten the battery load: the lighter each battery, the less weight a soldier

---

[42] Transcript, "Special Briefing on Objective Force Warrior and DOD Combat Feeding Program," 23 May 2002. Accessed 5 November 2006 at <http://www.globalsecurity.org/military/library/news/2002/05/mil-020523-dod03 htm>.

[43] MAJ Stuart Meyer, briefing slides, U.S. Army Infantry Center. Accessed 5 November 2006 at <www.dtic mil/ndia/2001smallarms/meyer.pdf>.

needs to carry. As important, though, is the total number of batteries a soldier must carry. The more quickly a battery is exhausted and the longer the mission, the more replacement batteries are needed. Recent research has produced batteries, such as the lithium ion family, that provide longer life and diminish soldier loads.

Large numbers of replacement batteries also mean excess cost. Whereas in training rechargeable batteries are used, in combat a used battery is most likely discarded. During Operation *Iraqi Freedom*, U.S. forces were at one point consuming 180,000 of one common type of disposable battery per month, at an acquisition cost (to say nothing of logistical burden) of almost $20 million a month.[44] Batteries that can be recharged in the field are a partial solution. Some portable solar panel rechargers are available. These rechargers offer weight savings for missions lasting longer than 24 hours and become cost efficient after 220 operational hours.

The Army, and the military at large, also suffers from a lack of standardization and the inability to cross-use batteries. In the 1970s and 1980s, the Army used 350 different types of 1.5-volt to 30-volt military-unique batteries.[45] This proliferation was both an operational and a logistical burden as well as source of great expense—in peacetime, in 1996, the Army spent $100 million on batteries.[46] By 2001, the number of battery types had been

---

[44] James Whiteker, Jason Hamilton, and Steven Sablan. *MBA Professional Report: Logistical Impact Study of Photovoltaic Power Converter Technology to the United States Army and the United States Marine Corps* (Monterey, CA: Naval Postgraduate School, December 2004), 2. The non-rechargeable battery examined in the study is the BA-5590, which, among other devices, powers the ground forces' primary tactical radios.
[45] "Army Battery Standardization: Rechargeable Batteries Power the Future Force." Defense Standardization Program Case Study, 2002, Accessed 7 January 2007 at <http://www.dsp.dla mil/cases.htm>.
[46] Ibid.

cut to 35. The Army's goal, though, is to get to 25 standardized battery types. This will lower acquisition costs and, through interoperability, bring operational efficiencies.

These power source-related issues—weight, efficiency, and standardization—all deserve continued Army S&T focus. As the Army proliferates portable high-technology devices throughout the force, and especially as it continues to engage in the kinds of conflicts that require extensive dismounted operations, S&T investment in improved power sources promises to have a pay-off across the force.

Moving Towards the Future: Network-Centric Operations Technologies

In this chapter we have discussed technologies that were and will continue to be crucial to the development of many weapons systems. These included both technologies and technical approaches that supported weapons development—M&S, HPC, and HSI—and technologies directly tied to fielded capabilities: gun technologies, night vision/infrared technologies, and power sources. We have noted their impact on the systems studied in the *Hindsight Revisited* reports and made it clear that S&T investment in these areas is likely to realize returns in more than one future system. Yet, while not diminishing the importance of these areas, it is essential to note a group of technologies that by their very definition offer even greater leverage on the capabilities of the Army and the military as a whole: those technologies that support network-centric operations. These technologies played only a small role in the *Hindsight Revisited* systems, but they will loom large in the development of future systems. M&S, night vision, and the like will continue to make important contributions to enhancing the capabilities of individual systems; network-centric operations will use information technologies to create a system of systems that will use the capabilities of individual platforms more effectively and efficiently than previously possible.

Network-centric operations are the military's response to the information age. Such operations pull together all elements of a joint force and through information technologies integrate units' situational awareness, knowledge, and capabilities to achieve rapid, decisive results. With shared, up-to-date situational awareness (of both Blue and Red forces) and the ability to communicate to adjacent units as well as to units at higher command levels, a netted force is able to increase the speed of command, adapt to changing circumstances, and deliver effects efficiently.

The military at large, and the Army in particular, is committed to realizing the potential combat benefits of operating in a network-centric manner. The concept is emphasized in both the 2005 National Defense Strategy and the 2005 Capstone Concept for Joint Operations and it is a central feature of FCS.[47] The FCS system of systems is referred to as 18+1+1, for 18 manned and unmanned platforms plus the soldier plus the network. FCS is founded in part on the idea that information conveyed by the network can be a significant force multiplier and can be substituted in some circumstances for mass. A netted force with an information advantage no longer needs to maneuver to contact: it can use remote sensors to find the enemy, share the information instantly, and respond before the enemy can act. An example illustrates the importance of the network to FCS: the Army is designing FCS combat vehicles to be lighter than some armored contemporary

---

[47] *Capstone Concept for Joint Operations*, Department of Defense, 2005. Accessed 17 January 2007 at <http://www.dtic.mil/ futurejointwarfare/concepts/approved_ccjov2.pdf>; *National Defense Strategy of the United States of America*, Department of Defense, March 2005. Accessed 17 January 2007 at <http://www.defenselink mil/news/Apr2005/d20050408strategy.pdf>.

combat vehicles because (it is hoped) the shared situational awareness created by the sensors and network of the FCS will allow its vehicles to either avoid sustained enemy fire or find and destroy the enemy first. If the Army cannot build and operate an effective, secure network, the capabilities of FCS will be significantly curtailed.

Though network-centric operations are enabled as well by complementary organizations and processes and highly trained people, their reliance on technology is plain. The concept will not work if the force cannot rapidly and effectively develop and share information. We touched on some such technologies in the *Hindsight Revisited* reports. The Abrams, for instance, was equipped with Force XXI Battle Command, Brigade and Below (FBCB2) systems that enabled vehicle commanders to track Blue forces. FBCB2 and a full spectrum of other C4ISR technologies will be crucial to enhancing the strength of the Army's weapons systems—strength derived in part from those technologies discussed earlier in the chapter—by enabling the Army to fight as a netted force and a system of systems. This requires technical work at the tactical, operational, and strategic level. The military's Global Information Grid (GIG) will allow the ready exchange of the full spectrum of defense information at the highest, joint echelons. The Army's piece of the GIG is called LandWarNet; it encompasses all elements of the service's communications architecture across all organizations and systems. Sensor platforms such as unmanned aerial vehicles (UAVs) are needed to populate the network with Red force information. Interoperable tactical radios and Blue force tracking systems must proliferate and they must be both vehicle borne and soldier-portable. Battle command systems are needed to manage and act on the flood of information.

The network-centric technology areas listed briefly above all exist in the force in some form, and even today the Army engages in network-centric operations. To realize the full potential benefit of the concept, however, further technical progress is needed. This is a crucial future challenge, if not *the* crucial future challenge, for the Army S&T program. S&T investment in such areas as M&S, ballistics, and sources of power can be leveraged for use across the Service, and investment in technologies for network-centric operations can multiply that leverage, enhancing the capabilities of the entire force.

In pursuing advances in the technical areas discussed in this chapter, the Army will make use of many sources. It will draw on the expertise of industry, universities and research centers, and our international partners. The Army will also turn to its in-house laboratory system. The *Hindsight Revisited* reports demonstrated the many significant contributions the in-house laboratories made to the development of past weapons systems. For the Army laboratories to make such a contribution in the future, the Army must pay close attention to critical areas of laboratory operations. The next chapter suggests some actions DOD and the Department of the Army can and should take to ensure laboratory excellence.

Chapter 4

Ensuring Excellence at the Army's Laboratories

A
s noted in Chapter II, the Army laboratories were critical to the development of past weapons systems. In our studies of four weapons systems in *Hindsight Revisited*, we found that in-house laboratories played an important role, providing, on average, about half of the CTEs. We also found that success in weapons technology depends on a close partnership between the in-house Army laboratories and those in the private sector, both of which are essential national resources. As noted in Chapter III, the need for excellence in Army S&T is as strong as ever. The Army has demanded and will continue to demand technical advances to meet emerging defense needs. In the current conflicts in the Middle East the Army laboratories are constantly providing new ways to protect our soldiers and enhance their effectiveness through technology. The success of the FCS program requires on a number of new, very advanced technologies, and the private-sector FCS LSIs are depending on the special expertise of the Army in-house staff. Both in-house and private laboratories must conduct their technical programs at the state-of-the-art and be pushing that art wherever possible.

Yet serious challenges confront the Army laboratories, and indeed all DOD laboratories. For a number of years efforts to reduce the federal budget and the size of the government have had a negative impact on the DOD laboratories, especially since the end of the Cold War. The size of the Army laboratories was reduced in the 1990s by almost half. S&T responsibilities have gradually shifted away from DOD and to the private sector. In this chapter we first discuss the role of the Army laboratories and then review the most pressing challenges, focusing primarily on personnel management issues since hiring and retaining top-flight research and engineering staff are critical for success. No amount of money or facilities can produce good results without expert personnel. We then examine several of the numerous studies of government laboratories in recent years that have addressed these challenges and review government attempts to improve laboratory operations. There have been some positive steps but much remains to be done. We conclude by discussing the current situation and suggesting improvements to the posture of the laboratories.

The Role of the Laboratories in Weapons Development

The Army laboratory system is comprised of those in the Army Materiel Command (AMC),[48] which were the principle laboratories in the *Hindsight Revisited* reports, the Army Corps of Engineers, and the Army Medical Command. These laboratories are managed day-to-day by their respective line military commands; however, for

---

[48] The Army Materiel Command's laboratories are: the Army Research Laboratory, the Aviation and Missile Research and Development Center, the Armaments Research and Development Center, the Communications and Electronics Research and Development Center, the Edgewood Chemical and Biological Center, the Natick Soldier Center, and the Tank-Automotive Research and Development Center.

program and budget management they come together in the office of the Army S&T Executive, under the Assistant Secretary of the Army for Acquisition, Logistics, and Technology.

Not everyone is familiar with the broad scope of the role that the Army laboratories play in developing and fielding weapons systems. A clear summary comes from the Federal Commission convened to look at the proposals for consolidation and conversion of defense laboratories as a part of the Base Realignment and Closure (BRAC) program of 1991.[49] The Commission, established by the Secretary of Defense, was drawn from government laboratory managers and senior people from outstanding private laboratories (Bell Laboratories and IBM, for example). Its report provides a clear, representative statement of the 10 main functions the DOD laboratories should perform:

- Infuse the art of the possible into military planning
- Act as principal agents in maintaining the technology base
- Avoid technological surprise and ensure technological innovation.
- Support the acquisition process
- Provide special-purpose facilities not practical for the private sector.
- Respond rapidly in time of urgent needs of national crisis
- Be a constructive advisor for Department directions and programs based on technical expertise
- Support the user in the application of emerging technology and introduction of new systems.

---

[49] *Federal Advisory Commission on Consolidation and Conversion of Defense Research and Development Laboratories,* Report to the Secretary of Defense, September 1991.

- Translate user needs into technology requirements for industry
- Serve as an S&T training ground for civilian and military acquisition personnel

To fulfill these functions, the laboratories must be active in four major areas: performing a full spectrum of research activities, from basic research to advanced development; supporting the PM offices; participating in the testing and evaluation of new weapons and technologies; and interfacing with the user.

The laboratories conduct basic research to provide fundamental understanding of problems related to new systems. Some of the work is done in-house and some in the private sector via grants and contracts. The Army Research Office (ARO) sponsors basic research in academia. This academic research is broad in scope and explores a wide spectrum of new areas that may be of interest to the Army in the long term. This will be discussed in more detail in the basic research section of this chapter.

All the Army laboratories conduct applied research and advanced technology development. Much of this is engineering work and the laboratories have very strong staffs to do this. As we found in the *Hindsight Revisited* reports, continuity in staffing at the in-house laboratories has led to a depth of expertise that is hard to come by elsewhere and would be very hard to recreate should the laboratories be shut down. The same applies to equipment and facilities. The technical staff works very closely with the PMs, industry, and, sometimes, other government laboratories, laboratories of our international partners, research institutes, and universities. Our studies of weapons systems have shown these relationships have been essential to the successful development of those four systems.

The in-house laboratories work closely with PM offices and, through them, industry. The laboratories work with the

PM offices in what is called matrix support. This refers to staff on the personnel rosters of the laboratories that are detailed to the PM. Matrix support positions may be temporary or relatively permanent. In some cases staff have moved all the way from ARL through an RDEC and a PM office and on to the contractor. This occurred in the development of the Abrams tank and proved to be a very effective method of improving communications as well as speeding technology transfer.

Another important role played by the Army laboratories is to support the testing and evaluation (T&E) community. The laboratories often perform informal evaluations of new ideas developed by contractors. The laboratories also often become directly involved in formal T&E and have designed and built some of the Army's test facilities. At APG and White Sands Missile Range, for example, ARL conducts test firings and other full-scale field evaluations.

An important advantage of having skilled and experienced technical staff at the in-house laboratories is that they are available to advise senior Army acquisition officials on the merits of technical proposals from outside the Army. Officials lacking the specialized knowledge required to evaluate the performance claims of such proposals can turn to the experts in the laboratories for unbiased reviews and recommendations. This has been called the "smart buyer" role of the laboratories, and it improves the likelihood that the Army will avoid costly mistakes.

As seen in Chapter II, the Army laboratories also perform the important function of involving the ultimate user—the soldier in the field—in the technology development process. The primary way the in-house laboratories accomplish this is by working closely with TRADOC and its schools and centers to receive definitions of Army needs, get advice on the formulation of technical programs to respond to these needs, and jointly evaluate the proposed solutions from the laboratories. Army S&T laboratories maintain user involvement by experimenting with their technologies and systems at TRADOC's 10 Army Battle Laboratories. The Battle Laboratories offer the Army S&T laboratories an opportunity to expose their technologies and systems to soldiers in simulated combat, combat support, or combat service support operations.[50] In most cases, the Battle Laboratories have linkages to other Services to support joint evaluations.

In the past, getting early user involvement through TRADOC was not an easy task, since the technology or system had to actually be built into a useful product for user assessment. With advances in M&S, as discussed in Chapter III, this has become easier. For example, with facilities such as the Fort Knox Unit of Action Maneuver Battle Laboratory using virtual simulations (vehicle simulators and live crews) to evaluate vehicle and crew performance, Army S&T laboratories can get warfighter performance data and feedback on FCS systems and components that have yet to be built.[51] The use of distributed, integrated, live, virtual, and constructive simulations provides a tremendous assessment capability to the laboratories and to TRADOC.

Army laboratories also have developed early user involvement through their Army Technology Objective (ATO) approval process. ATOs are established for technologies that the Army laboratories identify as key to developing a particular system or capability. As part of this process, TRADOC reviews the relevancy of ATOs to Army needs. The lead TRADOC organization for this process is the Army Capabilities Integration Center (ARCIC).

---

[50] John R. Wilson, Jr., "Battle Labs: What Are They, Where Are They Going?," *Acquisition Review Quarterly*, Winter 1996. Accessed 7 January 2007 at <http://www.dau mil/pubs/arq/94arq/wilso.pdf>.
[51] Unit of Action Maneuver Battle Laboratory web site. Accessed 5 November 2006 at <http://www.knox.army mil/center/uambl/>.

An additional path to the user is through the PMs. The PM offices work with the TRADOC schools and centers and hence provide the Army's scientists and engineers another way of interacting with those charged with representing the warfighters. We find that these close working relationships are essential to "getting it right" in developing new systems for the Army.

We now turn to the problems facing the laboratories that challenge their ability to carry out these roles.

Critical Issues Challenging the S&T Laboratories

The Army laboratories are confronted with significant obstacles to optimally performing the missions discussed above. In order to perform these functions, the 1991 Federal Advisory Commission held that an effective laboratory should have the following characteristics:[52]

- Clear and stable mission
- Highly competent and dedicated workforce
- Highly qualified and empowered leadership
- State-of-the-art equipment and facilities
- Close relations with the user/customer
- Strong basic research component
- Budget stability
- Champion in senior management above the laboratory
- Strong ties to other laboratories inside and outside the government

Management and budget challenges, however, are restricting the laboratories' abilities to maintain the excellence suggested by such characteristics. Issues are

---

[52] Taken in part from Federal Advisory Commission on Consolidation and Conversion of Defense Research and Development Laboratories, Report to the Secretary of Defense, September 1991.

largely in the following categories: hiring and retaining scientists and engineers; centralization of support services; failure of DOD to delegate to and empower senior laboratory managers to make their own decisions in managing their technical programs; overly close control of the technical work (micromanagement); and tight budgets that constrain staffing and reduce needed investment in equipment and facilities. We begin with a discussion of personnel issues that make hiring and retaining top-flight technical people in the laboratories difficult.

Personnel Issues

The most important factor for excellence in a laboratory is a top-flight technical staff. The laboratories should have outstanding research scientists and engineers in every aspect of the work. They should be supported by the best possible administrative and support personnel. The hiring and retaining of such outstanding people is difficult in any laboratory, but more so in the federal government. There are constraints that make hiring a slow process, place tight limits on salaries, and control the number of authorized positions and the number of high-grade positions. Laboratory management has had little authority over the hiring process; that has been controlled by the Office of Personnel Management (OPM) and DOD personnel offices at various levels above the laboratory management. So many layers of authority for personnel decisions are generally not found in the private sector. Thus private-sector laboratories can move quickly in hiring, sometimes making offers on the spot.

Competition for the best people is keen. The Army laboratories have an advantage in having a critical mission in defending the Nation and in their generally good to excellent equipment and facilities. These advantages are offset by lower salaries and salary gaps at the top grades, the low enthusiasm that young people have for the federal government, and the appeal of private-sector opportunities. Furthermore, the number of U.S. citizens enrolling in S&T programs in universities has not kept up with the demand.[53] Many students graduating in physical science and engineering nonetheless opt for careers in business or health. Though plenty of non-citizen graduates are coming into the market, the military laboratories normally cannot hire non-citizens without encountering a lot of red tape. The movement of foreign nationals has been further restricted since the September 11 terrorist attacks. Finding and hiring strong candidates requires that the laboratories have aggressive recruiting efforts that go far beyond the simple posting of vacancy announcements.

Leadership. Excellent leadership of the Army's technical work should ensure that the staff performs as well as possible. Poor leadership would likely reduce the staff's ability to perform. Therefore, the laboratories should have, at all management levels, people well-qualified in management in general and in management of S&T in particular. This generally means that managers should have a technical education and personal experience in performing laboratory work. They also should know how to administer the laboratories. Sometimes such leaders are found within, and sometimes outside, the laboratories. For senior positions such as laboratory director there should be a comprehensive search for the best candidates.

Further, there have been numerous efforts to develop a cadre of military scientific officers without noticeable success. In years past the laboratories had many young officers spending a year or two working technical problems and becoming skilled in the art of S&T. Some of these soldiers held advanced degrees in science or engineering.

---

[53] Michael Marshall, Timothy Coffey, Fred Saalfeld, and Rita Colwell, "The Science and Engineering Workforce and National Security," *Defense Horizons 39* (Washington, DC: Center for Technology and National Security Policy, National Defense University, April 2004).

However, after the end of the Cold War force reductions virtually eliminated the possibility of training young officers in this way. Today very few uniformed military are in the Army laboratories, and very few mid-career officers have the necessary technical experience to effectively oversee the Army's technical programs. As a result, most senior officers in the Army acquisition process are not experienced in R&D. They understand only poorly the need for appropriate budgets and staffing.

Other Management Challenges

In addition to the aforementioned difficulties in handling personnel at the laboratories, there are challenges in funding, management practices, capitalizing facilities and equipment, and maintaining excellence in basic research. These challenges are outlined below.

Budgets. Federal budgets are noted for their instability. They are usually not enacted on time so that the government must operate on continuing resolutions. This means that proposed programs may not be initiated until, often, a full quarter after the start of the fiscal year. The laboratories' budget proposals are altered by the layers of management above them in DOD and then reviewed again by the four committees (two in authorizations and two in appropriations) in Congress. In recent years the budget has contained a substantial numbers of "earmarks," designations as to where the funds should go without benefit of the usual scrutiny of authorization and appropriations hearings. Since the budgets often are not increased to accommodate the new projects, existing work must be reduced to pay the bill.

Even more fundamentally, budgets rise and fall with the fortunes of the Nation in terms of security and the economy, falling when the Nation's situation seems secure and booming when it is threatened. For the laboratories, the boom usually does not mean more support for forward-looking research but rather funding for using current knowledge in support of the warfighters in the field. The laboratory staff do not decry this effect; rather the staff feel stimulated to be providing such help to those in harm's way. Nonetheless, the programs aimed at long range Army goals are constrained. During busts, hiring is either frozen or restricted. Travel accounts are reduced so that most staff cannot attend scientific meetings. Faced with such uncertainty in the support by the Administration and Congress, potential recruits to the laboratories often turn away. To illustrate, the severe cuts in budget and staffing after the end of the Cold War affected the laboratories a great deal In the Army laboratories both budgets and staff levels were reduced between 40 and 50 percent.[54]

Management. In the past 15 years DOD has centralized such functions as financial management, personnel management, and purchasing and has pressed local managers to contract out many other support functions. In one recent case DOD proposed that even the alteration of individual laboratory space be controlled not by the laboratory director but by a central support facility not under the laboratory's control. In such an environment it is difficult to maintain a high level of research output. These constraints make the position of laboratory director frustrating and less appealing to new candidates. It does not need to be so. For instance, in a report on the development of the sabot for the M829A2 projectile (an advance discussed in the *Hindsight Revisited* Abrams report), researchers specifically cited the lack of micromanagement

---

[54] Edward A. Brown, *Reinventing Government Research and Development: A Report on Management Initiatives and Reinvention Efforts at the Army Research Laboratory,* ARL-SR-57, Army Research Laboratory, Adelphi, MD, August 1998.

as important to the success of the program.[55] There are examples today of federal agencies where laboratories are managed extremely well and generally do not face the aforementioned difficulties, e.g., the National Institute of Standards and Technology.

Facilities. Partially offsetting the inhibitions on attracting talented personnel are the laboratories' generally good equipment and facilities. The present facilities at the laboratories were not come by easily. Building up the facilities has been done over many years. For example, most of the facilities for research in ballistics have been constructed, maintained, and upgraded since the 1930s. These facilities have received continuing investment so as to keep up with advances in the science and engineering of ballistics work. A case in point is the special-purpose enclosed test firing range for handling armor targets and penetrators containing DU. DOD also has invested heavily in high end computers and the Army has benefited from this. At APG there is the HPC MSRC mentioned in Chapter III, and the Corps of Engineers Waterways Experiment Station at Vicksburg also has an MSRC. In addition, as noted in the *Hindsight Revisited* reports, the Army has been fortunate in being able to share large scale experimental facilities at three NASA technical installations: NASA Langley, NASA Glenn, and NASA Ames. In all likelihood the Army never could have justified building its own wind tunnels and the like.

Ironically, the Army has realized some benefits from turmoil surrounding the BRAC process. When a laboratory was closed and the staff relocated, new buildings and replacement equipment was provided. For example, when the Watertown Arsenal was closed in BRACs 1989 and 1991, the program was transferred to APG. A new, state-of-

---

[55] Bruce P. Burns, *An Executive Summary of the M829A2 Sabot Technology Program*, ARL-TR-350, Army Research Laboratory, Adelphi, MD, February 1994.

the-art building was constructed and furnished with appropriate equipment. Similarly, when the electronics S&T program at Fort Monmouth was transferred in BRAC 1991, the Army constructed a new research building at Adelphi, MD. It has world-class clean rooms and other up-to-date equipment needed for research at the sub-micron level.

Support for Basic Research. Another significant challenge is to develop appreciation among military leaders for much of the Army's basic research program. At one time it was generally accepted in the technical community at large that a strong research laboratory (or a strong commercial enterprise depending on technology) needed to have a certain amount of basic research—the number often cited is 15 percent or so of the base S&T budget. While most of that work in the Army has been motivated directly by technical challenges in the applied work, some should be driven by the desire to expand the technical horizons of the laboratory. In the Army of today the majority (~80 percent) of the 6.1 funds are held by ARL, of which almost two thirds goes to universities.[56] The Army's RDECs are budgeted to do relatively little basic work; the Army Corps of Engineers and the Army Medical Command have a modest amount of basic research funding. These groups could benefit greatly from an increased basic research budget. (See p.88 for more on basic research.)

Past Studies of Defense Laboratories and Responses

In light of the above challenges, groups within and outside the Government have sought corrective actions. In this section, we review the most important of these studies, the majority of which relate to the key issue of personnel

---

[56] Emails to authors from the Office of the Deputy Assistant Secretary of the Army for Research and Technology, November, 2006.

management, and indicate the nature of the responses by Congress and the Executive branch.

Past Studies of Defense Laboratories

The laboratories have been studied repeatedly. Coffey et al. estimate that about 100 studies of Government laboratories have been done since 1962.[57] The very fact that so many studies have been done underscores the importance of the laboratories. The motivation behind these studies arose, on the one hand, from a sense that the private sector could simply take over the laboratories' role in acquisition. Many in the private sector criticized the performance of the in-house laboratories and essentially said "we can do it better." On the other hand, others asserted that the in-house laboratories are essential and steps should be taken to ensure quality and responsiveness. The following review of just a few of the more recent studies reveals a clear consensus on both the qualities the laboratories should have and steps that should be taken to ensure that the laboratories meet those goals. In particular, the studies emphasize the importance of changing the laboratories' personnel management system.

In 1983 the White House Science Council published what has become known as the first Packard report.[58] This report recommended a number of steps to strengthen the federal laboratories; among them were peer review of the work and empowerment of the laboratory directors in all aspects of laboratory management. One of the examples cited in the Packard report was the experiment in personnel

---

[57] Timothy Coffey, Kenneth Lackie, and Michael Marshall, "Alternative Governance: A Tool for Military Laboratory Reform," *Defense Horizons 34* (Washington, DC: Center for Technology and National Security Policy, National Defense University, November 2003).

[58] *Report of the White House Science Council Federal Laboratory Review Panel* (Washington DC: The White House, May 1983).

management being conducted at the Navy's Surface Weapons Center at China Lake, CA, and the Naval Ocean Systems Center at San Diego. Known as the China Lake alternative personnel system, this project, set up in 1980, was the first DOD personnel demonstration program as authorized under the 1978 Civil Service Reform Act. This program included a stream-lined classification system (a few broad pay bands) and a pay system based on performance rather than longevity.

The Packard report recommended that all federal laboratories try something similar. President Reagan accepted the report and directed the federal departments to implement it. The White House Office of Science and Technology Policy (OSTP) drafted model legislation to enable the laboratories to proceed (for a synopsis of the model legislation see Appendix A).[59] However, OPM and the personnel offices in the cabinet departments resisted and the legislation was not introduced. (The House Science Committee, however, became interested and drafted a similar authorization for the National Bureau of Standards (NBS, now the National Institute of Standards and Technology, or NIST). The bill became law and was implemented beginning in 1988. Please see Appendix B for further details.)

The Defense Science Board (DSB) has conducted a number of studies of the DOD laboratories.[60] Its 1987 study led to the establishment in 1989 of the Laboratory

---

[59] *Progress Report on Implementing the Recommendations of the White House Science Council's Federal Laboratory Review Panel, Vol. 1— Summary Report* (Washington, DC: Office of Science and Technology Policy, Executive Office of the President, July 1984).

[60] *Report on the 1987 Summer Study on Technology Base Management* (Washington, DC: Department of Defense, December 1987); *Defense Science Board Task Force on Defense Laboratory Management, Interim Report* (Washington, DC: Department of Defense, April 1994); *Defense Science Board Summer Study on Defense Science and Technology* (Washington, DC: Department of Defense, May 2002).

Demonstration Program (LDP). (LDP was later succeeded by the Laboratory Quality Improvement Program (LQIP), which was then folded into Vice President Gore's National Performance Reinvention program.) In his memo implementing the 1987 DSB recommendations, Deputy Defense Secretary Atwood said; "Recent studies . . . have shown that the productivity and effectiveness of the DOD laboratories can be significantly improved by implementing specific changes in procedures involving personnel management, research-related contracting, facilities refurbishment, and management authority of technical directors."[61] In the personnel management area, the laboratories and centers in the LDP sought further delegations of authority. The requests included the authority to directly hire personnel, set salary rates, approve performance-award amounts consistent with the activity's budget, approve recruitment and retention bonuses, and eliminate time-in-grade requirements for promotion.

Unfortunately, most of the personnel-related authorities desired by the laboratories and centers under the LDP required the formal approval of OPM under the provisions of the Civil Service Reform Act of 1978 (CSRA). Despite lengthy discussions, OPM remained unwilling to utilize its CSRA authority to establish any additional personnel demonstrations for Defense laboratories. OPM did not want simply to duplicate features tested in "China Lake" and LDP, but wanted features that could explore additional concepts that might apply across the government. OPM insisted on cost neutrality, and wanted extensive justification for why any variation from standard personnel practices was needed for the laboratories. Without OPM action, little progress resulted.

---

[61] *Science and Technology Community in Crisis*, Naval Research Advisory Committee Report (Washington, DC: Office of the Assistant Secretary of the Navy (Research, Development, and Acquisition), May 2002), 18.

In Section 913 of the National Defense Authorization Act (NDAA) for FY2000 (see Table 3), Congress proposed that the DDR&E use university study teams to look at the past relevance of defense laboratories and evaluate the laboratories' current work and its utility in the future. The historical part of this undertaking was done by two groups at George Washington University,[62] the current and future portion by a team of experienced managers convened by NDU.[63] The retrospective look at the laboratories contains the following remark: "DOD laboratories have been carrying out relevant work. The case studies [in the retrospective review] also built a cumulative case for the value to the country of having a diverse, capable system of DOD laboratories, acting as a focused part of the U.S. national research enterprise." The NDU team issued four reports on different technical areas and aspects of laboratory management. The studies by the NDU team found that relevant work is being performed and the laboratories are well-focused on the military's technical needs. The team expressed some concern that some laboratories may be too relevant; that is, that their work is too concentrated on short-term needs and not enough on long-term opportunities. They worried about the difficulties

---

[62] *The Contributions of Department of Defense Laboratories to U.S. Warfighting Capabilities Case Studies of Twelve Laboratories* (Washington, DC: Center for International Science and Technology Policy and Security Policy Studies Program, The George Washington University, August 2002).

[63] *Report #1: Sensors Science and Technology and the Department of Defense Laboratories* (Washington, DC: Center for Technology and National Security Policy, National Defense University, March 2002); *Report #2: Information Science and Technology and the Department of Defense Laboratories,* July 2002; *Report #3: Weapons Science and Technology and the Department of Defense Laboratories,* December 2002; *Report #4: A Study of the Connectivity Between the Defense Laboratories, Industry, and Academia in the Area of Information Technology,* July 2003.

caused by DOD personnel practices and they urged more R&D programs explicitly directed at joint service needs.

In 2001, the office of the DDR&E requested the Naval Research Advisory Committee (NRAC) to conduct a study of the health of the three Service "corporate laboratories"— ARL, the Naval Research Laboratory (NRL), and the Air Force Research Laboratory (AFRL). [64] The tri-Service panel consisted of high-level persons drawn from non-DOD government laboratories, academia, and industry. It concluded that DOD laboratories have an essential future role to play, listed characteristics of a world class laboratory (similar to those listed by the Federal Advisory Commission on Consolidation and Conversion report cited earlier[65]), and wrote that previous studies were "mostly well done, but few of their recommendations [have been] implemented." The report's recommendations were first that the DDR&E seek and obtain the commitment of the Secretary of Defense and the Service Secretaries as to "the need for and the importance and value of the Service Corporate Research Laboratories by demonstrating continuing support"; second, that the Secretary of Defense utilize existing authorities to establish a separate personnel system for the scientists and engineers in the Services' corporate laboratories; and third, that the DDR&E "develop and propose to Congress additional legislation that would enable the Services to experiment with alternative governance structures." The report explicitly identified the issues that should be on the list: "salary caps, burdensome procedures, inability to renew facilities and equipment, lack

---

[64] *Science and Technology Community in Crisis.* The NRAC study report, prepared with the participation of the Army Science Board and the Air Force Scientific Advisory Board, included a detailed summary of many of the earlier studies.

[65] *Federal Advisory Commission on Consolidation and Conversion of Defense Research and Development Laboratories,* Report to the Secretary of Defense, September 1991.

of laboratory director authority, and poor support services." Although focused only on the three Service research laboratories, these recommendations could easily apply to all of the AMC laboratories (ARL and the RDECs), the Corps of Engineers laboratories, and the laboratories of the Army Medical Command.

The material covered in the NRAC report has recently been expanded upon and brought up to date in a recent report by Kavetsky et al.[66] While directed at the Navy, this report's ten recommendations should be considered by the Army. Several of the recommendations pertain to improving the ability to hire and retain top-flight technical staff. The report also recommends an increase in funding for Navy S&T in general (to three percent of total obligation authority) and of special funding for Navy research exploring the frontiers of knowledge. The three percent figure has been proposed in many external studies and endorsed in both Senate and House committee reports. The FY07 Army budget shows S&T (6.1, 6.2, 6.3) at 1.5 percent of the total of $111.8 billion. An increase to three percent would be difficult to obtain at any time but is especially hard to achieve now with heavy pressure on the budget from the costs of the conflicts in the Middle East.

Legislative and Executive Branch Actions
Attempts to Strengthen LDP/LQIP. Concerns about the Government laboratories, and DOD laboratories in particular, are well documented. As discussed above, in the years since the creation of LDP, the U.S. Congress has addressed the issues surrounding the Defense laboratories a number of times, and the executive branch has taken some action. In addition to the statutes cited above there have been several pieces of legislation that authorized

---

[66] Robert Kavetsky, Michael Marshall, and Davinder Anand, *From Science to Seapower* (College Park, MD: Center for Energetic Concepts Development Series, University of Maryland, 2006).

laboratory-related reforms or further studies of laboratory problems. Progress, however, has been halting.

In early 1993 the three Service Acquisition Executives committed to an intensive joint effort to get LDP back on track. This initiative was subsequently chartered as LQIP and essentially succeeded the LDP. At about the same time, the National Performance Review (NPR) was getting off the ground, as a part of the Vice President Al Gore's initiative to "reinvent Government." (The NPR was subsequently renamed the National Partnership for Reinventing Government.) Soon after, the DDR&E requested that the LQIP laboratories be designated as a single "reinvention laboratory" under the aegis of the NPR. This request was approved in March 1994; it authorized the laboratories and centers to request waivers of Service and Office of the Secretary of Defense (OSD) policies and regulations in support of DOD's "Installations 2000" vision, which called for "Installation commanders empowered with the responsibility, authority, flexibility, and resources to make the requisite decisions based on what is best for client tenants and their respective missions." Based on submissions from the Services, DDR&E initially included one Defense, two Navy, four Army, and four Air Force laboratories in the NPR program. Ultimately, the Army extended its participation to all of its R&D laboratories.

One of the basic tenets of the NPR was the identification and elimination of barriers to efficiency, either through legislative change or by waiving burdensome and unnecessary regulations. In fact, during the early days of the NPR, the Secretary of Defense (SECDEF) required the senior leadership of the Department to take action on all waiver requests within 30 days or see the waiver take effect automatically. Disapprovals of waiver requests could only be based on statutory or national security considerations. In the end, however, execution of this SECDEF directive was spotty at best.[67]

Numerous reforms were attempted under the new LQIP/NPR umbrella. As noted above, the 1987 DSB study had recommended expansion of the China Lake demonstrations to all DOD laboratories and centers. Such an expansion was not possible under the 1978 Civil Service Reform Act (CSRA), so new legislative authority was required. The LDP/LQIP devoted several years to securing acceptance of the concepts in this proposal within the Pentagon and OPM, but were never able to get a draft bill approved for submission to Congress. Eventually, Congress itself took action, citing the DSB and other studies for justification. Section 342 of the FY1995 NDAA authorized the SECDEF, with the approval of OPM, to carry out personnel demonstration projects "generally similar" to the China Lake project at DOD laboratories designated by SECDEF as S&T reinvention laboratories.

In August and September 1995, the DDR&E, acting on the authority granted to the SECDEF by the FY1995 NDAA, requested five waivers from the personnel office of OSD:

- the establishment of a definition of a high grade in a pay-band system,
- redefinition of the term, "Full Term Equivalent" (FTE),
- removal of high grade controls,
- prohibition of hiring freezes, and
- exemption from the Priority Placement Program for internal placements.[68]

---

[67] Brown; *Science and Technology Community in Crisis*; W.C. McCorkle, J.R. Houston, J.A. Montgomery, L. McFawn memorandum to W.S. Rees, Chair, Laboratory Quality Enhancement Program in the Office of the DDR&E, "Authorities Necessary to Effectively Manage the Defense In-House Laboratories," 21 August 2006.

The Assistant Secretary of Defense for Force Management Policy (ASD (FMP)) declined to approve them and placed a 60-day hold on a decision, pending negotiations to resolve differences. Ten months of difficult negotiations between (1) DDR&E and the Services (acting for the LQIP) and (2) ASD (FMP), the Office of Management and Budget, and OPM (as stake-holders for the existing system) followed. Finally, in June 1996, the Deputy Secretary of Defense approved a negotiated settlement covering:

- a definition of high grades in pay-banding systems;
- authorization to exceed FTE authorizations by two percent on a temporary basis to cover Intergovernmental Personnel Act assignees, experts and consultants, students, and faculty working at LQIP laboratories and centers;
- a nine-month grace period to complete staffing actions at LQIP laboratories and centers after the imposition of a hiring freeze, and
- a process for resolving qualification determinations under the Priority Placement Program.

Although these presented only marginal changes to the system, they were quite simply the best deal that could be obtained. In the end, they provided little benefit to the LDP/ LQIP laboratories and centers.[69]

The National Security Personnel System (NSPS). NSPS, a DOD-wide personnel overhaul, is the centerpiece of the Pentagon's most recent efforts to address personnel problems across the entire DOD. This new personnel system, authorized in the NDAA for FY2004, establishes a

---

[68] Kenneth Lackie, former scientific staff assistant to Technical Director, Office of Naval Research, email to authors 12 December 2006.
[69] Ibid.

few wide pay bands in place of the former 15-step General Schedule and adopts the principle of pay for performance.

The authorities offered by NSPS do not approach those enjoyed under LDP/ LQIP, much less those additional authorities recommended by various studies and requested by the DDR&E. The main characteristic of LDP/LQIP is the empowerment of the laboratory management; the program allows laboratory managers to waive many statutes and use provisions of 60 other statutes, powers not available in NSPS. NSPS would move most of the decision authority to higher levels in the bureaucracy. Table 2 below summarizes some of the differences between the LDP/LQIP authorities and those under the NSPS.

Table 2[70]

Comparison of Elements of LDP/LQIP vs. NSPS

LDP	NSPS
Can waive many parts of Title 5, the Civil Service System	NSPS cannot waive these items
SECDEF approves changes to LDP	OPM must approve changes in NSPS
May pay starting salaries anywhere in a pay band	Limited to 30 percent above minimum
LDP has a Pay Band V for senior positions	NSPS does not have such a band
Supervisors not automatically paid more than group members	Assumes supervisors are automatically superior in pay

---

[70] McCorkle et al, "Authorities Necessary to Effectively Manage the Defense In-House Laboratories."

82

Promotions from band to band without competition	Crossing pay bands requires competition
Manage most of Human Resources (HR) function	HR functions held at levels above the laboratories
Classification, recruiting, qualification, and hiring authorities reside with laboratory managers	Classification and related authorities are held at levels above the laboratories

In approving the NSPS in NDAA for FY2004, Congress exempted the defense laboratories until October 2008, so that adjustments and changes might be made. Congress has revisited the personnel issues in Defense authorization measures for both FY2005 and 2006, and indicated some concern over problems left unsolved and about the impact on the defense laboratories of the new one-size-fits-all NSPS. The FY2005 NDAA says, in Sec. 1107, that the "the Under Secretary of Defense for Acquisition, Technology, and Logistics and the Under Secretary for Defense for Personnel and Readiness shall jointly develop a plan for the effective utilization of the personnel management authorities" already enacted for increasing the effectiveness of the defense laboratories. This internal study was never conducted. In Sec. 1123 of NDAA for FY2006 Congress said that "the Secretary of Defense shall commission an independent study to identify the features of successful personnel management systems of the highly technical and scientific workforces of the Department of Defense and similar scientific facilities and institutions." The Act asks for "a comparative assessment of the effectiveness of the Department of Defense technical workforce management authorities and practices with that of other similar entities." Congress also pointed out earlier authorities granted in FY1995 and 1999 and sought information on the status of the use of these authorities.

In response to these expressions of congressional intent the DOD pointed to a 2004 study it had sponsored on best practices in managing large R&D organizations. The study looked at current practices in the private sector and defense laboratories and focused on laboratory quality, personnel, and management.[71] In it, the in-house laboratory managers discussed what constitutes quality in the laboratory, including the importance of having top quality personnel. They also expressed concerns about funding stability, priority ranking processes, the trend toward centralizing or regionalizing support services at the expense of flexibility and performance at the laboratories, the balance in investment in-house and contracting out, and the proper ratio of basic and applied research vice product development. Apart from this study, no other assessment has been conducted in response to Sec.1107 of the FY05 NDAA and Sec.1123 of the FY06 NDAA.

Table 3
Some Recent Legislative Actions Regarding DOD Personnel

Legislation	Description
Civil Service Reform Act, 1978	Authorized alternate personnel systems.
NDAA for FY1995, Sec. 342	Made permanent the Navy's China Lake demonstration system
NDAA for FY2000, Sec. 913	Requested studies of capabilities of DOD laboratories—past and future

---

[71] Gerald Krueger and Edward Molnar, *Exemplary Practices in Management of Large Research and Development Laboratories,* The Wexford Group International, Vienna, VA, September 2004.

NDAA for FY2001, Sec. 1114	Delegated OPM authorities to the Sec Def
NDAA for FY2004, Sec. 9902	Authorized the NSPS for DOD
NDAA for FY2004, Sec.9902c	Exempted LDP/LQIP laboratories from NSPS until 2008
NDAA for FY2005, Sec. 1107; and for FY2006, Sec. 1223	Requested studies of the NSPS vs. the DOD LDP/LQIP; also a study comparing these systems with those at the best private laboratories

Areas for Action

The weight of some 20 years of studies and experience strongly suggests that the laboratories need a marked increase in their ability to manage their own affairs, especially in personnel. Recent legislative and executive branch actions have not sufficed. Steps to increase flexibility in managing laboratory personnel should be taken within the current laboratory governance framework. More attention should be paid to ensuring a strong basic research component and to opportunities to collaborate with other parties.

Personnel

As described in most of the major laboratory studies to date, personnel-related issues are the foremost problem confronting the laboratories. DOD should approve the request recently put forward by senior laboratory managers from each of the Services to the DDR&E.[72] The Services asked that the laboratories should be allowed to manage their resources to budget and not face controls on staffing levels. The additional ceilings on various categories of personnel ought to be eliminated and DOD should restore to the laboratories the management of their principal support services. If approval for all of the laboratories' personnel management proposals is granted it would empower the laboratory directors in much the same fashion as in most of the best private and government laboratories. Table 4 highlights some of these additional needs in personnel management.

Table 4

Additional Needs in Personnel Management
(Features not now in LDP/LQIP and not planned
for NSPS[73])

- Extend LDP/LQIP features uniformly to all LDP/LQIP laboratories
- Approval for more laboratories to join LDP/LQIP
- Extend delay of transferring LDP/LQIP into NSPS from 2008 to 2014
- Drop controls of number of positions in Pay Band V
- Raise pay ceiling to that of Senior Technical and top Senior Executive Service positions and in, some extraordinary cases, to Executive Level II
- Establish a Senior Scientific Service under the DDR&E

---

[72] W.C. McCorkle et al, "Authorities Necessary to Effectively Manage the Defense In-House Laboratories."
[73] Ibid.

- Modify Voluntary Early Retirement Authority/Voluntary Separation Incentive Pay to apply to specific positions and individuals
- Temporary hiring authority similar to that under the Intergovernmental Personnel Exchange Act
- Use of personal service contracts for clerical and secretarial positions

These items are directed at reducing the amount of top-level management of areas that the laboratories feel would best be delegated to the laboratory directors

Personnel issues also include the need to pay increased attention to the method of selecting people for key leadership positions in the laboratories. True national searches should be mandated by OSD and conducted at the operating levels to locate the very best talent for positions such as laboratory director and directors at the next lower echelon. Such laboratory senior managers should be, as a prerequisite, educated at the graduate level in science or engineering and experienced both in conducting S&T themselves and in managing complex S&T organizations.

Governance

There have been some proposals to change the governance structure of the DOD laboratories. The alternatives ranged from the current status—the government-owned, government-operated (GOGO) model—to government-owned, contractor operated (GOCO), e.g., the DOE Laboratories; independent government-chartered corporation such as the Tennessee Valley Authority; and outright privatization. A 2003 study by Coffey et al analyzed these alternatives and suggested that further consideration be given to a government-chartered corporation (GCC) model.[74] Among other things, it would give laboratory management wide latitude to address some of the personnel issues raised elsewhere in this chapter.

In our opinion, however, the current structure (GOGO) is still the best option. Despite its faults, it keeps the laboratories very much in the mainstream of Army activities and assures them of close everyday contact with their immediate customers in the acquisition community as well as with their ultimate clients, the warfighters. The advantage of this positioning is apparent in the efforts of the laboratories to solve the special problems arising in Iraq and Afghanistan. The rapid transition of new technology from the laboratory bench to the warfighter in times of crisis has been facilitated by having the laboratories within the Army. The idea of moving the laboratories outside the normal bureaucracy arises from frustration over the inability of the current management structure to support and strengthen properly these important resources. However, the best solution to this problem is for the current leadership to confront the challenges faced by the in-house laboratories head-on (and implement the recommendations made in this book) rather going down a GOCO or GCC path, where the strong coupling with the user would be jeopardized.

Keeping the laboratories inside the Army does not mean that the thrust of the S&T portfolio should not continually evolve. Change is inherent in the S&T enterprise. To illustrate, the Army is now (2007) sponsoring a study at the National Research Council on how the Service should address network S&T. This study was stimulated in part by the important role the subject is playing in the FCS program as well as the closing of Fort

---

[74] Coffey et al, "Alternative Governance: A Tool for Military Laboratory Reform."

Monmouth and the transfer of the Communications and Electronics Research, Development, and Engineering Center to APG. This transfer presents an opportunity to create a focal point on networking capabilities and the possibility of hiring new and different staff (it is estimated that a majority of the current Fort Monmouth personnel will not make the move).[75] The NRC study will consider the competencies needed and the size of the various component programs. The study is also to address the use of new partnerships to share resources and expertise with industry.[76]

A Strong Basic Research Component

Basic research is important to laboratory programs in many ways. Such research may be curiosity- or problem-driven; i.e., the investigator may be motivated by a desire to expand knowledge in general, or by the needs of a particular project or program area. We have pointed out earlier that the applied work on particular systems rests on prior basic studies often performed many years earlier. Most of the basic research pursued by the military is problem-driven—either in today's work or in areas of potential interest in the future. In addition, during the course of applied and development work serious barriers arise that can only be overcome by carrying out additional basic research studies. When such problems do arise it is usually too late to go out and hire personnel to do the needed basic work.

Thus, basic research is directly involved in weapons development. But there are many additional benefits to

---

[75] Estimate based on experience with 1991 BRAC. Michael Marshall, email to authors, 27 June 2005.

[76] *Statement of Task for the Network Science, Technologies, and Experimentation Center,* Board on Army Science and Technology, National Research Council, The National Academies, Washington, DC, 2006.

having basic research as an integral part of an S&T development program. Having within the laboratory basic work at the cutting edge will help ensure that the more applied work is also at the frontier. Absent this, the applied work may drift away from the frontier and become second-rate. Discussion with colleagues performing basic research—at technical conferences and through informal everyday contacts in the corridors and lunch rooms—keeps all the staff informed of the latest developments in science and engineering. Another advantage is the attraction of basic work to new graduates. For a new scientist or engineer with an advanced degree, basic research more nearly resembles the master's or doctoral thesis work he or she has just completed. This eases the transition from the university. Recruiting is thus facilitated. Also, the staff members performing basic research are likely to be closely involved with researchers in universities, National Laboratories, and the like, again helping to make sure that the laboratory is fully current in its fields of concentration. It is often possible to make use of the knowledge developed in these external laboratories.

Collaboration

Finally, as the nature of R&D in the world is changing, so too should the laboratories be changing their modes of operation. It would especially benefit the Army to adapt its S&T program to capitalize on private-sector expertise in those research areas in which industry has a substantial technical advantage. Information technologies, broadly speaking, are an example of a discipline in which the commercial market has produced and will continue to produce advances at a greater rate than the government can hope to match by itself.

Fortunately, more and more, collaboration among entities outside the parent laboratory organization is the norm. Researchers team up with colleagues around the world, laboratories execute agreements to collaborate with research facilities elsewhere, and research is done on computers linked over the internet—the more so as M&S becomes more prevalent. The laboratory that is not collaborating in some fashion with its peers in the public and private sectors is more likely to be parochial and, ultimately, to become second-rate. For the Army laboratories, which have always worked closely with their contractors' laboratories, this trend means that they should have working relationships with centers of excellence in universities and research institutes both domestic and foreign, as well as other government laboratories.

Recently, with 6.1 monies, the Army has created large Centers of Excellence in areas of interest to the Army for the purpose of putting more emphasis on certain technology areas and promoting collaboration between industry and in-house laboratories. Some of these Centers are: the Institute for Creative Technology at the University of Southern California (mentioned in Chapter III), the Institute for Collaborative Biotechnologies at the University of California at Santa Barbara, and the Institute for Soldier Nanotechnologies at the Massachusetts Institute of Technology. These Centers are consortia of universities and private companies; they are led by universities and funded with research grants from ARO. Another set of consortia, funded by ARL with 6.1 and 6.2 monies and led by industry, are more tightly coupled to the in-house laboratories. These are called Collaborative Technology Alliances. They are managed by ARL; the programs are planned by teams of people from ARL and the alliance participants, and personnel are exchanged between the ARL and alliance members. These exchanges facilitate information flow back and forth and make technology transfer into the Army much easier.

We have detailed in this chapter several areas where the Army can and should strengthen its S&T programs. It seems to us that these actions have already been authorized, should be relatively easy to implement, and would greatly enhance S&T laboratory operations. It would be a shame not to accept these challenges.

Chapter 5

Concluding Remarks and the Way Ahead

The foundation for this book came from the *Project Hindsight Revisited* studies, in-depth retrospective looks at four of the Army's key weapons systems: the Abrams main battle tank, the Apache attack helicopter, and the Stinger and Javelin man-portable missiles. Our approach was based on a 1969 DOD report titled *Project Hindsight*. Our studies focused on establishing CTEs for each system. All told we uncovered 135 of these key technology events for the four systems and analyzed them as to their origin, their funding, the role of management both at the laboratory and program managerial level, and the extent of public and private-sector collaboration.

This book has drawn on the *Hindsight Revisited* studies' key findings, and on two of them in particular. First, though the four systems had quite different capabilities, certain technologies were important to all of them. Such technologies, which have the potential for broad impact across the force, deserve particular attention from the Army S&T program. We have singled out several technologies that enhance the S&T program's capabilities and several technologies that enhance the capabilities of forces in the field. In the former category, we have focused on M&S, HSI, and HPC. In the latter category, we have addressed gun technologies, night vision technologies, and batteries. Though this book only illustrated a small slice of their technical potential, it is clear that Army and DOD investment in these (and other such) areas could be leveraged across multiple future weapons systems.

This book has also focused on the health of the Army laboratory system. The *Hindsight Revisited* studies revealed that the Army's in-house laboratories and industry were between them responsible for the great majority of the CTEs. Collaboration between the two, facilitated by the PM offices and sometimes supplemented by contributions from other government laboratories, universities, and some of our overseas partners, was absolutely essential to the successful development of the studied systems. While all these parties were important, the Army leadership must look to ensure the continued strength of the entities firmly under its control, the in-house laboratories.

While the in-house laboratories face several challenges (among them, declining budgets and a diminished focus on basic research), perhaps the most important is the difficulty in hiring and retaining talented personnel. No amount of money or facilities can produce good results without expert personnel. Our survey of important personnel-related studies and legislation has found that though there is broad agreement on the importance of personnel reform and on some of the steps that should be taken, the federal government has consistently withheld from laboratory managers authorities that would improve the laboratories' personnel posture.

The health of the laboratory system should be of acute concern to DOD and the Army. As a starting point, it is important to monitor and assess laboratory operations, both to ensure that the laboratories are meeting the Nation's needs and to clearly highlight those areas (such as the personnel management practices discussed above) that would benefit from adjustment. Drawing on published studies and their own experiences in both the government and private sector in performing and managing R&D, Lyons et al. made several proposals on this issue.[77] They listed means to measure the effectiveness of a laboratory and suggested a three-pronged approach: external peer review for quality, customer rating of timeliness and effectiveness, and review by senior Army stakeholders as to the overall S&T approach and priorities. Some progress is being made. For example, ARL uses the National Research Council to do a peer review, the technical directors of the RDECs do the customer ratings, and a special team of senior Army managers at the three-star level has at one time represented the stakeholders. Unfortunately this team of stakeholders has been disbanded. We believe something equivalent to this review process should be initiated at all S&T laboratories.

Assessments are only a beginning. Preserving and promoting a robust Army capability in weapons development will require the combined efforts of many people, chief among them the Secretary of Defense, the Secretary of the Army, and the congressional committees. These parties need not look far for places to begin. We recommend three broad steps. First, the Secretary of Defense and the Service Secretaries should articulate their continued commitment to the importance and value of the Service research laboratories. The DOD should start by making full use of the authorities already granted it in law, such as relaxing personnel ceilings and relying on budget controls. The DDR&E should develop and propose to Congress legislation to address additional laboratory issues, particularly legislation that will enhance the authority of the

---

[77] John Lyons, Joseph Mait, and Dennis Schmidt, "A Strategy for Improving the Army Research and Development Laboratories," *Defense and Technology Paper 12* (Washington, DC: Center for Technology and National Security Policy, National Defense University, March 2005).

laboratory directors. These, in fact, were the recommendations of the NRAC report discussed in Chapter IV.[78] They should apply to all the DOD laboratories, not just Army laboratories and not just the corporate laboratories (NRL, ARL, and AFRL).

The *Hindsight Revisited* reports also showed the value of close collaboration between the Army technical staff and industry. Whatever future laboratory reforms are undertaken, careful attention should be paid to promoting such collaboration. The Army should undertake more long-term collaborations with the private sector, especially in basic and early applied research in areas where the private sector has taken the lead. When using LSIs, efforts should be made to ensure that the in-house expertise continues to be utilized; reassuringly, this seems to be happening with the FCS program.

Without champions at the very top of the Executive Branch it is unlikely that further improvements will be made, and it is likely that the posture of the laboratories will decline. Indeed, the recent record shows that even when Congress has shown its interest in addressing these problems and enacted authorization to do so, the Executive Branch has resisted changes. An example of another approach to laboratory reform is the legislation setting up an alternative personnel system at NBS.[79] Patterned on the model legislation drafted by OSTP, Congress gave the authority to create the personnel system directly to the Director of NBS, thereby circumventing some levels of the Executive branch.[80] This legislation outlined in great detail the key provisions the new system should contain; it enabled the NBS to plan and obtain the system they wanted

---

[78] *Science and Technology Community in Crisis.*
[79] Sec. 10, Public Law 99-574, 28 October, 1986. This was a five year demonstration project; it was extended indefinitely in Public Law 104-113, 7 March, 1996.
[80] See Appendix A.

as long as it was consistent with its authorization act. (See Appendix B for some details.) Such an approach, in which Congress by-passed OPM and other levels of bureaucracy, may be necessary for the DOD laboratories.

We believe that the *Hindsight Revisited* studies of the role of technology in weapons development have been useful and we suggest that the other Services undertake similar retrospective reviews. For the Army (and this very likely applies to the other Services) we conclude that for future success in developing weapons systems there must be a balance of technical competences in the DOD laboratories and their counterpart laboratories in industry. Neither group can do the job alone. The Army in-house laboratories must be supported by sufficient funding, strong leadership, and top flight technical staff. They require champions at the topmost levels of the Army and of DOD. We have offered here suggestions as to the measures required to maintain and strengthen the laboratories. We believe strongly these steps must be taken to insure the effectiveness and the safety of the Nation's warfighters in the future.

Afterword

Historical studies of the impact of DOD laboratories on military system development, as exemplified by *Project Hindsight Revisited*, have invariably demonstrated essential contributions from the laboratories in technology development, responsiveness to military crises, and above all, support to the acquisition process. In a technologically complex world it is essential that the government maintains the expertise to oversee the development of complex systems. As this report points out, and what I have observed from a Naval perspective, is that DOD laboratories have for many years served as a S&T training ground for civilian and military personnel and many of these personnel have subsequently moved to technology management and technology acquisition.

Despite the outstanding track record of the laboratories there is currently a mounting concern that the DOD's laboratory capabilities, particularly in the oversight of very large, complex systems development, have been reduced by a decades long process of realignments, base closings, and the perception that there is more challenging work for researchers in the private sector. Other dynamics have also conspired to change the roles of the laboratories in military system development. These include the break-up of the Soviet Union, which left the United States without a military technology competitor; the consolidation of the military industrial base into a few dominant "system" houses or prime contractors; and the shift in developmental funding, particularly in information technology, from the DOD to industry.

These trends have led the government to explore the concept of the Lead Systems Integrator (LSI) for major development projects such as the Army's Future Combat Systems (FCS). The LSI concept has the prime contractor assuming many roles in program development, such as system architecture development and program management, that previously were done by government personnel. What is being found is that while the DOD can outsource much of its technology capabilities to industry it still needs to maintain the expertise essential for program oversight and adherence to military requirements. The lack of knowledgeable oversight by government personnel has recently become of concern to Congress in some major LSI-directed programs, including FCS. At the heart of the issue is that an LSI is ultimately responsible to its shareholders and profit margin while the government alone is responsible for the security (including economic security) of the Nation.

The solution to strong government technology oversight lies in strong DOD laboratories that serve the government's role in technology development and technology insertion but are not technology competitors with the industrial base. Fortunately, despite draw-downs, the DOD laboratories retain a strong core competency in military-related research and development. Furthermore, many of the base realignments have tied the laboratories closer to the system acquisition commands. It is vital that the essential functions of the laboratories be maintained and strengthened so that they may continue to provide needed oversight and sound acquisition analysis for the government in a time of rapidly advancing technology and changing national threats.

Dr. Eli Zimet
Distinguished Research Fellow
National Defense University
Former Head, Expeditionary
Warfare Department, Office of
Naval Research

OSTP Response to Personnel Recommendations of the Federal Laboratory Review Panel

The following information is from the OSTP progress report[81] on implementing, in the personnel area, recommendations of the White House Science Council's Federal Laboratory Review Panel.[82]

OSTP proposed draft model personnel legislation. The key features are:

- Permits agencies to establish alternative personnel systems;
- Applies to scientific and technical personnel in the Federal laboratories;
- Bases pay on performance rather than on longevity;

---

[81] *Progress Report on Implementing the Recommendations of the White House Science Council's Federal Laboratory Review Panel.*
[82] *Report of the White House Science Council Federal Laboratory Review Panel.*

- Creates broad pay bands;
- Simplifies job classification;
- Allows the pay cap to be waived for up to five percent of the covered positions . . . limited to specially qualified scientific and technical personnel;
- Permits the agency head to classify positions and fix compensation so as to make positions competitive . . . [with] comparable positions outside of the Federal government;
- Allows the inclusion in the alternate personnel systems of positions now in the Senior Executive Service; and
- Permits the Naval Weapons Center and Naval Ocean Systems Center personnel systems (i.e., China Lake) to become permanent.

In addition to the model legislation, the Progress Report goes on to list personnel actions within existing authority to improve the laboratories. The interagency group under the OSTP Committee on Federal Laboratories recommended the following:

- Exclude laboratories from the current [1983] proposal to reduce the number of employees in the Civil Service grades 11 through 15;
- Allocate to the laboratories more positions for specially qualified scientific and technical personnel under provisions of 5 USC 3104 and allow some supervisory responsibility for such positions [these are the ST positions];
- Provide government-operated laboratories with blanket permanent direct hire authority for all professional scientific and technical positions in the laboratories;

- Provide government-operated laboratories with blanket direct hire Excepted Service Appointment Authority for research associates;
- Exempt laboratory summer hires in technical positions from manpower ceilings;
- Include special rate schedules for engineers and other manpower shortage occupations in annual cost of living adjustments that are applied to other Federal pay scales; and
- Increase the probationary period for scientific and technical personnel from one year to a more suitable period, such as three years.

Appendix B
New Personnel Authorities at the NBS/NIST

The following are provisions of the authority for a personnel system granted by statute to the director of the National Bureau of Standards (NBS), now the National Institute of Standards and Technology (NIST).

- Aggressive recruitment based on improved authorities. Direct hiring authority at the discretion of the NBS director. Flexible entry salaries.
- Classification into broad pay bands (NBS chose to use five, in four career paths: Scientific and Engineering; Scientific and Engineering Technician; Administrative; and Support). Classification and hiring decisions to be done by line managers with advice and counsel from NBS personnel specialists.
- Pay comparability. NBS was directed to produce studies of pay in comparable positions in the private sector. Based on these studies, the NBS was directed to make adjustments to the pay of all of its employees so as to prevent the differential from increasing, and, if funds are available, to raise the

salary of all employees to reduce the pay gap with the private sector.

- Supervisory and managerial pay differentials are included.
- Pay for performance. There are five levels of performance. The amount of pay increase depends on the rating; poor performers get no merit increase and the lowest (unsatisfactory) performers do not receive a cost of living raise. For individuals well rated but at the top of the pay band, performance bonuses are given.
- NBS chose to maintain budget neutrality in its compensation system; i.e., the total costs should not increase over what it would have been under the standard Civil Service system.
- Hiring and retention bonuses.
- Reduction-in-force. Each career path is a separate competitive area; i.e., "bumping rights" no longer extend from one career path to another.
- Senior Executive Service and Senior Technical positions are not included in the NBS system.
- The probation period for new employees may be extended to three years for employees on career-conditional appointments.

Many of the above provisions have since been added to other personnel demonstrations. However, some (e.g. pay comparability, direct hire authority, supervisory pay differentials) have not been approved in DOD.

Appendix C
The *Hindsight Revisited* Reports

Summary of CTEs in the Development of the Abrams Tank

The following summary provides brief explanations of the CTEs discussed in the Abrams *Hindsight Revisited* report. Introductory material, conclusions, discussion of any advances not dubbed CTEs, and some analysis have been omitted. This summary also dispenses with references and citations. The full report, with complete footnotes, can be found at: <http://www.ndu.edu/ctnsp/Def_Tech/ DTP22%20Critical%20Technology%20Developments%20 Abrams%20Tank.pdf >.

Armament Related CTEs

Main Gun

- The Army wanted a larger gun than the 105mm with which the Abrams was originally equipped. Though the likely candidate, a 120mm smoothbore gun under development in the United States and West Germany, was deemed not ready at the time, the M1 designers were directed to provide a gun mount and turret that could handle the larger and heavier weapon when it became available **[CTE 1]**. This forethought paid off in 1981, when the Army elected to adopt the 120mm smooth bore cannon. The Army purchased the gun design and the know-how and equipment to manufacture it from West Germany. Watervliet Arsenal installed a rotary forge, worked out the remaining manufacturing difficulties, and started production of the gun.

- Work at the Benet Laboratory improved the resistance of large caliber guns like the 120mm to unexpected brittle fast fracture. Using linear elastic fracture mechanics to assess gun tube failure, researchers developed an understanding of the fatigue and fracture behavior of the tubes **[CTE 2]**. This work developed a new fracture toughness test specimen from a thick-walled gun barrel and a new test procedure that made use of the specimen. With these developments, both resistance to fast fracture and the fatigue crack growth rate could be measured using a cost-effective test methodology. This design information was used in conjunction with a manufacturing process known as autofrettage to increase gun tube life **[CTE 3]**. Around 1970, Watervliet Arsenal developed swage autofrettage, in which pressure applied by a mandrel replaces the hydraulic pressure. This process, now used world-wide, was first used on the M68 105mm gun and subsequently on all 120mm guns.

Gun Accuracy

- Errors in aiming that were tolerable for close-in combat were unacceptable for long-range firing. In the 1960s, researchers at Frankford Arsenal put together an error budget (an analysis of the sources and size of the errors) for tank guns **[CTE 4]**. Their report was the basis for early fire-control systems fabricated at the Arsenal.

- The first step in addressing these errors was to find a way to measure the effects. Researchers at the Ballistics Research Laboratory (BRL) were able to develop experimental, computational, analytical, and statistical models that could determine the relative magnitude of each component of error **[CTE 5]**. This resulted in the development of a multi-disciplinary approach that created integrated high-level physics-based models of the system.

- Analysis showed that lack of straightness of the gun barrel was one of the key sources of error. A special machine press was designed at Watervliet Arsenal to address tube straightness and profile after manufacture and to correct tubes brought in from the fleet for overhaul **[CTE 6]**.

Penetrators

- In developing the 120mm gun, the Army made the key technical decision that it should be optimized for long-rod penetrators **[CTE 7]**. The Army funded a large research effort at BRL to perfect an armor-piercing, fin-stabilized, discarding sabot (APFSDS) round for the 120mm gun **[CTE 8]**. This was important in the development of one of the mainstays of the Abrams' armament, the M829 series of APFSDS long-rod penetrators.

- Modeling work was done at BRL on the structure of the penetrator in an effort to craft a long rod with the highest possible ratios of length to diameter and also with the highest possible density [CTE 9]. BRL did systems analysis of interior, exterior, and terminal ballistics for the rods. Based on this modeling, researchers determined the optimal materials for use in the round.

- Depleted uranium (DU) replaced tungsten as the main penetrator material in the mid 1970s [CTE 10]. DU has superior ballistic qualities: the tip of a DU penetrator shears such that it remains sharp as it passes through armor, so that even with diminished velocity a DU round can defeat a target's armor. Battelle Northwest Laboratory suggested a new process to improve the compressive strength of the DU rod [CTE 11]. The Oak Ridge Y12 plant helped out as well by supplying tungsten and DU for the program.

- Research work on penetrators drew on unique and vital Army resources [CTE 12]. Supercomputers enabled ballisticians to run very complex models on penetrator-target interactions. Another facility, Experimental Facility 9 at Aberdeen Proving Ground (APG), was constructed to handle DU in bullets and armor.

Sabots

- Modern kinetic energy tank ammunition is normally composed of a narrow long-rod penetrator surrounded by a sabot, which expands the diameter of the round to the full barrel diameter of the gun. A slipping obturator was developed at Picatinny

Arsenal and used to make APFSDS ammunition compatible with the rifled 105mm gun **[CTE 13]**. The slipping made it possible for the sabot to engage the rifling of the barrel while the round itself turned at fewer revolutions per second than does the obturator.

- A tipping ring was designed for the rear of the sabot that pivoted the sabot segments so that they were in a position to fly cleanly away from the penetrator **[CTE 14]**. R&D work at Picatinny Arsenal and BRL also produced a series of designs for the sabot focusing on exact shape of the scoops or ramps. The final double ramp sabot shape used on the M829 was the result of computer modeling by BRL.

- Sabots were first made of steel, then of aluminum and magnesium alloys, and today are made of composite materials. Staff at BRL used finite element stress analysis to establish the proper geometric design of optimized, minimum-mass, aluminum sabots for both the 120mm M829 and M829A1 **[CTE 15]**. The next step was the incorporation of composite materials. Careful experimentation defined the lay-up of the fibers in the composite. Investigators at BRL—teamed with composite material specialists at the Lawrence Livermore National Laboratory and with industry—developed the means to design the architecture and processing for the first composite sabot used for the M829A2 round **[CTE 16]**.

Propellants

- Among the most important aspects of developing better propellants was modeling. Efforts at BRL resulted in an improved computer code called the XKTC **[CTE 17]**. XKTC has been widely distributed among America's allies. Studies with a scanning electron microscope revealed the micro-mechanisms of the behavior of the propellant in the M829 and other 120mm rounds. These studies led BRL to develop a new propellant design that incorporated partial transverse cuts at regular longitudinal distances in the sticks instead of a lengthwise slot, thus allowing multiple perforations in long, large-diameter sticks that both packed well and provided highly progressive gas generation rates with time **[CTE 18]**. This combination of features provided an increase in interior ballistic performance.

Armor and Other Survivability Related CTEs

Armor

- Through the use of M&S at the U.S. Army Tank-Automotive Research and Development Center, and other system engineering tools, the designs of the M1 series tanks optimized the vehicle silhouette, the location and type of armor, the location of the crew and vulnerable components, and many other survivability factors **[CTE 19]**. A new hull configuration demanded innovations in the production process. In-house engineers from several laboratories and arsenals established joint design and welding techniques to be used in fabricating the hull **[CTE 20]**.

- The M1 program started with the intent to use what the Army called "Special Armor" developed through exchanges with the United Kingdom and through U.S. indigenous advances. The British concept, also known as Chobham armor, complemented parallel work in U.S. Army laboratories **[CTE 21]**. Details of the Abrams' armor design and composition are classified, but this much can be said: instead of using a single material—steel—the Chobham concept uses steel over one or more layers of different materials, each layer designed to perform a different function against incoming munitions **[CTE 22]**. The armor is therefore a layered composite. Developers made heavy use of experiments and early computer models to develop ever more complex and effective composite armor.

- BRL researchers developed new armor concepts, the most notable being one that incorporated DU **[CTE 23]**. Selected for its high density and special performance in high-shear fracture, DU makes an ideal armor component. This upgrade was fielded on the M1A1 and M1A2 models.

- Staff at BRL designed a separate compartment to stow ammunition; the compartment makes use of automatic doors **[CTE 24]**. The compartment design also provided sufficient venting for any explosion by installing blow-out panels that would direct energy from the blast away from the crew.

- Controlling the interaction of stowed high explosive warheads was critically important to an ammunition compartment concept. BRL developed less shock- and crush-sensitive warheads and included plastic shields between stowed rounds **[CTE 25]**. Researchers also developed a new test rig, a three ton pendulum that quantifies the response of ammunition components to impact **[CTE 26]**.

- When the 120mm gun technology was purchased from Germany, a design was included for combustible sidewalls in the casings. However, Army tests revealed some deficiencies, such as incomplete combustion, low strength, and trouble with surface coatings. A team from Picatinny Arsenal and two contractors—Honeywell and Armtec Defense Products (now Esterline Armtec)— resolved these difficulties **[CTE 27]**.

- Infrared sensors in the Abrams' crew compartment can detect a fire in a few milliseconds and (if it is a petroleum fire) extinguish it within tenths of a second using Halon agent **[CTE 28]**. This upgraded fire suppression capability grew out of a TACOM program that funded efforts in the private sector and academia that significantly advanced the state of the art in key areas such as sensor technology and flow of the suppression agent.

- Another area of crew protection for the Abrams was defense against nuclear, biological, or chemical (NBC) attack. The Army's Edgewood Chemical Biological Center (ECBC) was tasked with analyzing the feasibility of a hybrid collective protection system that would make use of both individual gear and compartment-wide filtration and overpressure. ECBC awarded a contract to develop Hybrid Collective Protection Equipment (HCPE). This HCPE effort included the development of the NBC filter currently being used on the Abrams **[CTE 29]**.

System Testing, Modeling, and Analysis

- BRL began developing models and computer codes to predict the vulnerability of combat vehicles (both aircraft and ground vehicles) shortly after WWII **[CTE 30]**. Until 1984, these models were deterministic, meaning that they could not account for the stochastic nature of the interaction between an attacking munition and the target vehicle.

- Beginning in 1984, BRL developed computer models to predict the outcome of specific live-fire test shots **[CTE 31]**. These new models characterize behind-armor debris by the spatial distribution, mass, and velocity of all fragments and predict which components or crew members will be hit by the behind-armor fragments. The models take into account the probabalistic nature of the input variables and produce probability distributions for the output variables.

Engine and Drive System CTEs

Engine

- Though some felt that the Abrams should have a diesel engine (an existing program had generated a possible model), a gas turbine engine was chosen for the tank. The AGT1500 gas turbine engine was part of the Chrysler design that won the XM1 competition **[CTE 32]**. A viable gas turbine engine was available in part because TACOM initiated an R&D program around 1960 in gas turbine engine technology for ground vehicle applications **[CTE 33]**.

- Engine designers had to find a way to provide sufficient air cleaner volume to allow the vehicle to operate for a reasonable period without cleaning the air filters. The upgraded M1A2 filtration system, developed and supplied by Donaldson Company, uses a pulse jet air cleaner that removes the need for the crew to manually clean the filter **[CTE 34]**.

Transmission

- Allison Transmission, a division of General Motors, created a hybrid hydrostatic-mechanical transmission **[CTE 35]**. Allison ultimately recommended that the hydromechanical technology not be pursued for the XM1 effort.

- Allison developed a transmission based on the Allison on-highway commercial transmission design. The new transmission, the X1100, is a 35-cubic-inch, radial displacement, hydrostatic steering mechanism that remains unique in the world. **[CTE 36]**. The X1100 also incorporated power-assisted service brakes that allowed maximum-effort brake stops with minimal brake pedal force. Continued development of the X1100 multi-plate, wet disk brake system has resulted in an extremely durable, reliable, high-performance brake system **[CTE 37]**.

Track and Suspension System

- With the increased speed and weight of the Abrams, the suspension system needed attention, especially if the tank was to perform well on rough terrain. The suspension system for the Abrams M1A2 has larger torsion bars and larger shock absorbers, which double the damping capacity **[CTE 38]**. To produce a robust torsion bar it was necessary to use steel in heavy section processed to high strength levels. This required the use of advanced heat-treating methods available in the private sector.

- The T–156 track system initially installed on the M1 had integral rubber track pads that exhibited far less track life than designers had hoped. A TACOM-sponsored re-engineering effort resulted in the T–158 track system, which had replaceable pads that met the initial track-life goals **[CTE 39]**. T–158 track system was later optimized for weight reduction to produce the T–158LL track system, which was installed on the M1A2.

- Research was performed in academia to analyze a problem with the drive sprocket on the M1A2. The result of this effort was a new sprocket better suited to the high tension. This resulted in a 20-percent improvement in track end connector wear and a reduced acoustic signature **[CTE 40]**.

Vetronics, C4ISR, and Fire-Control CTEs

Vetronics

- TACOM envisioned a significant change to the electronic integration of the Abrams using a digital architecture approach, dubbed vetronics, similar to the integrated avionics used in aircraft. General Dynamics Land Systems, working under a DARPA/TACOM contract, generated important algorithms that were applied to vetronics. This formed the basis of the M1A2's digital intravehicle architecture. The vetronics architecture supports almost instantaneous, push-button selection on the gunner's controls, provides connections to significantly more inputs from a variety of sensors (range, cant, wind, etc.), and allows more rapid adjustment of the position of the main gun [CTE 41].

C4ISR

- An early-1980s concept put forth by the Army Science Board was aimed at tying together the various pieces of information generated during the course of ground vehicle operations to give tank crews the ability to know where they are, to see what other tanks see, and to exchange information between vehicles [CTE 42]. The Army laboratories at TACOM, Picatinny Arsenal, and CECOM moved forward using digital capability to make enhanced battlefield communication a reality. The concept of an overarching Battlefield Management System was first manifest in the Intervehicular Information System (IVIS) [CTE 43]. Communications networking architecture and protocols were developed by CECOM for IVIS in the late 1980s and early 1990s.

- A key enabler of this battlefield communication concept is the M1A2's Position/Navigation (PosNav) system **[CTE 44]**. PosNav is a nuclear-hardened, autonomous navigation system with a GPS component backed up by an inertial navigation system. When connected to IVIS using the protocols described above, the system enables the creation, sharing, and constant updating of the battlefield picture.

- C4 capability was further enhanced by the Vehicle Intercommunication System **[CTE 45]**. This system included the capability to reduce ambient vehicle noise significantly by using an electronic Active Noise Reduction unit.

- The idea of extending C4 capabilities across all battlefield elements was expanded upon in FBCB2 packages that were developed by CECOM in the late 1980s and early 1990s **[CTE 46]**. These packages were appliquéd on the M1A1 and embedded into the M1A2SEP. FBCB2 significantly enhanced situational awareness as well as command and control to the lowest echelons.

Thermal Imagery

- Non-standardized components meant that the Army bore the cost of outfitting and maintaining a variety of thermal-imaging systems for individual weapons platforms. A common sensor module approach, pioneered by the Night Vision Laboratory (NVL) in the early 1970s, reduced the skyrocketing cost of custom imaging systems **[CTE 47]**. These Common Modules were the basis for a whole generation of thermal imagers that the Army used not only in the Abrams, but also in missile systems and helicopters. As a result of these efforts, the Army and the other Services were able to achieve parts standardization and significantly reduce costs. This approach became accepted production practice by industry and has led to affordable first and second generation FLIR thermal imaging systems on the Abrams.

- An NVL researcher and an industry researcher quantified recognition performance in terms of performance related to a minimum resolvable temperature difference **[CTE 48]**. This measurement technique was subsequently standardized at NVL and has become an industry standard for performance evaluation of infrared systems.

- The M1A2 added a commander's independent thermal imaging system, which also was fabricated using the common modules approach **[CTE 49]**. This system gave the tank commander a sighting system completely independent of the gunner, thus allowing the commander and the gunner to identify and track separate targets simultaneously.

Fire-Control System and Related Sensors

- The Abrams' lethality is distinguished by its digital fire-control system, which integrates data collected by sensors that gather any information that might affect the flight of the round and processes it with a digital ballistic computer **[CTE 50]**.

- The ruby laser, the first successful optical laser, was the basis for the first type-classified, man-portable rangefinder in the mid-1960s. In the 1970s, Army researchers first developed laser rangefinders based on Neodymium: Yttrium Aluminum Garnet (Nd: YAG) lasers **[CTE 51]**. Additional efforts yielded an industry-developed eyesafe Erbium laser rangefinder, which replaced the Nd: YAG laser on the Abrams **[CTE 52]**.

- The muzzle reference system (MRS) measures the deviation of the muzzle with respect to a fixed point on the turret **[CTE 53]**. This design was modified when Benet technologists discovered severe resonances that greatly amplify local strains and accelerations in the tube—enough to cause damage to the MRS **[CTE 54]**.

- Information from the laser rangefinder, the muzzle reference system, and other sensors are collected and provided to the other primary component of the fire-control system, the Abrams' digital ballistic computer. In the 1970s, engineers at Frankford Arsenal were able to develop a digitized fire computer with greatly improved speed and accuracy **[CTE 55]**. It represents a major leap forward from the analog system used by the M60A1.

Summary of CTEs in the Development of the Apache Helicopter

The following summary provides brief explanations of the CTEs discussed in the Apache *Hindsight Revisited* report. Introductory material, conclusions, discussion of any advances not dubbed CTEs, and some analysis have been omitted. This summary also dispenses with references and citations. The full report, with complete footnotes, can be found at: <http://www.ndu.edu/ctnsp/Def_Tech/ DTP%2026%20Apache%20Helicopter.pdf>.

Engine and Powertrain

T700 Engine

- Problems with insufficient power, poor reliability, high maintenance needs, and poor specific fuel consumption in existing helicopter engines led the Army to develop a new engine for its next generation of helicopters. The Aviation Applied Technology Directorate (AATD) was tasked to set requirements for these engines, with special emphasis on maintainability **[CTE 1]**. Their approach, drawing on their "1500 Demonstrator Engine" program, was to estimate the outer limits of the existing technology and incorporate challenging goals in the request for proposals.

- General Electric's (GE's) design for a 1500-1600 horsepower engine was selected to power the Army's new generation of helicopters, including both the Apache and the Black Hawk. GE simplified the combustor by going to a machined-ring configuration that was made in one piece, improving performance and life **[CTE 2]**. GE also developed a scheme for removing particles from the incoming air, using cyclonic effects (swirl vanes) in the front of the compressor together with an air pump to eject the separated particles **[CTE 3]**.

- GE engineers also overcame several initial challenges. Forced vibration of compressor blades required some engine redesign **[CTE 4]**. Also, the unusually high rotational speed of the compressor shaft and the power shaft called for improvements in the properties and design of the disks and blades. In one of the first applications of powder metallurgy, GE learned how to manufacture these assemblies in one piece that they termed "blisks," using powder metallurgy with a nickel-based alloy **[CTE 5]**.

- The Army engineers co-located at NASA Glenn in Cleveland developed a process, now standard in the industry, for applying ceramic coatings to line the combustor and the blades in the hot section of the engine **[CTE 6]**. Ceramic coating allows higher operating temperatures and hence greater efficiencies.

- The Army funded 6.1 basic research at universities on rare-earth magnets that enabled significant weight and size reductions for starters and generators for the Apache **[CTE 7]**.

- A significant DOD program paralleling and reinforcing the Apache work is the Joint Turbine Advanced Gas Generator program **[CTE 8]**. It is a three-phase program with ambitious goals to produce a 120 percent increase in the ratio of shaft horsepower to engine weight and a decrease in specific fuel consumption of 40 percent.

Transmission

- Among other key features, the Army required that the transmission be protected against sudden loss of oil so that the Apache could sustain significant damage and still fly home. The resulting run-dry design enables the transmission to operate for up to 30 minutes after losing oil **[CTE 9]**.

- For gears and bearings, the Army/NASA work overcame barriers to higher transmission performance in terms of speed, loading, and operating temperatures. Work on double-vacuum melted, high hot-hardness bearing steel and on gear alloys "doubled the power density compared to previous engines and vastly improved reliability" and removed a major barrier to progress in engine and transmission technology **[CTE 10]**.

Survivability and Structural Advances

Vulnerability and Susceptibility Reduction

- Important strides were made in ballistic protection. The Army Materials Laboratory at Watertown developed the concept of using a transparent laminate armor material to separate front and rear cockpits **[CTE 11]**. The material, a glass/polycarbonate laminate, was patented used above the seat line.

- The development of ceramic composite materials by industry and the Army laboratory at Watertown led to additional ballistic protection for Apache crew members **[CTE 12]**. Boron carbide was demonstrated to be the most effective ceramic material; when coupled with glass-reinforced composite back-up material, it was able to defeat small arms threats. Later, the use of Kevlar was proposed by Watertown as a material for the rear of the boron carbide armor. This combination provided even better weight efficiency and was chosen for use in the Apache for the helicopter crew seats.

- Another flight-safety effort addressed the hazards presented by communication and power lines. A cooperative AATD/Canadian effort developed a system for wire strike protection **[CTE 13]**. The design was very basic: a cutter/deflector system consisting of an upper and lower cutter that would sever steel, copper and aluminum wires. The resulting system was retrofitted to the Apaches in the mid-1980s and built into most Army helicopters.

- The Army placed great emphasis on protecting the Apache's crew during a hard landing **[CTE 14]**. This emphasis was especially prudent considering that the Army required the Advanced Attack Helicopter (AAH) to execute nap-of-earth (NOE) operations; low-speed, ground-hugging flying, sometimes at night, increases the probability of having an accident. An AATD helicopter crash research program involving subject matter experts from private-sector safety groups was initiated in the late 1950s and continued into the 1980s. Data from the program led to the first ever crash-survival design guide for light fixed- and rotary-wing aircraft **[CTE 15]**. As a result of this significant step forward, the Apache crew has a 95 percent chance of surviving a crash with no spinal or thermal injury when the helicopter approaches the ground at rates of up to 42 feet per second.

- The Army's development of sound design criteria for crashworthy fuel systems produced the largest payoff in the endeavor to improve rotorcraft (including Apache) crash safety. Changes were made in both fuel system materials and design **[CTE 16]**. Self-sealing and tear-resistant polymers enhanced protection of the fuel tank. Changes were also made in the breakaway, self-sealing valves; fuel lines; filling system; and vents.

- The Apache also incorporated several advances designed to prevent the aircraft from being targeted at all **[CTE 17]**. In-house laboratories such as AATD and CECOM, working closely with industry, provided advances in the following areas: engine exhaust IR suppressor design; the ALQ-144 IR countermeasure system, a device designed to jam incoming heat-seeking missiles; low-signature paint and windshield concepts; and radar and laser warning systems.

Structural Advances

- During the Apache's development, the idea of using integral armor for structural load-carrying purposes as well as for ballistic protection was advanced by Hughes Helicopter during discussions with Army materials scientists. Hughes utilized this innovative approach to realize significant weight savings **[CTE 18]**.

- Among the next major structural challenges for ongoing Apache development work was the design of an all-composite rotor blade for the aircraft. Efforts by Hughes, Kaman Aircraft, and AATD have paved the way for the eventual addition of an all-composite rotor blade to the Apache **[CTE 19]**. When the Apache was first fielded, composite materials were not yet available that could meet the AAH program ballistic requirement at an acceptable weight. Research done after initial fielding of the Apache by AATD, combined with experience gained by Boeing in developing composite blades for other aircraft, resulted in new materials and analytical tools for blade spar sizing. This allowed the design of a new single spar, all-composite main rotor blade for the Apache, which is planned for fielding in 2008 **[CTE 20]**.

- A group of Army researchers co-located at NASA Langley has been among the leaders in quantifying the delamination process in composite materials. In the 1990s, this work led to accepted test methods and guidelines for determination of delamination, which is highly important to the ability to predict failure in composite materials **[CTE 21]**.

Avionics, Fire Control, and Weapons

TADS/PNVS

- The Target Acquisition and Designation Sight and the Pilot Night Vision Sensor (TADS/PNVS) are key to the Apache's ability to operate at all times, especially in all conditions, to find, track, and fire upon targets. Designed by Martin Marietta,

TADS/PNVS is an integrated system that pulled together for the first time thermal imaging devices, laser range finders and laser designators **[CTE 22]**. The sensors for both units are located in a rotating turret mounted on the Apache's nose.

- Critical to the TADS/PNVS is the FLIR sensor. With work done for the Abrams tank program on a minimum resolvable temperature metrics as a foundation, NVL produced a mathematical model that could be used to evaluate the performance of IR devices **[CTE 23]**.

- The Apache required significantly longer operational target engagement ranges than existing systems; it demanded a FLIR with more than twice the resolution of the one used by the Abrams. The Apache initially had a non-standard high line rate imaging system that was enabled by significant improvements in detector and optics quality sponsored by NVL and Frankford Arsenal **[CTE 24]**. The FLIR in PNVS also had special needs night piloting and NOE flying. The different trade-offs in FLIR system design compared to those for targeting systems were pursued to optimize sensor design for piloting parameters, such as field-of-view and sensitivity sufficient for background discrimination **[CTE 25]**.

- The Apache crew uses the Integrated Helmet and Display Sight System, better known as IHADSS, to slave the helmet to the TADS/PNVS sensor turret so that the pilot views the terrain and target area through the PNVS **[CTE 26]**. Similarly, TADS is slaved to the co-pilot/gunner's helmet, as is the chain gun. Infrared linkages developed by industry and Frankford Arsenal make this possible. With IHADSS, symbols representing such important information as aircraft heading, air speed, altitude, and target cueing are superimposed on displays that fit in front of the pilot's and gunner's right eyes **[CTE 27]**. This enables the crew to fly and fight without having to look at the panel-mounted displays in the cockpit. The Army laboratory co-located at NASA Ames played a key role in optimizing symbology for the Apache. Design standards were established and transitioned to industry to ensure that the proposed symbols and the pilot's feel for the motion of the helicopter were matched **[CTE 28]**.

Fire Control

- Important work on fire control for Army rotorcraft was done at BRL during the 1970s and early 1980s. BRL research provided a general 6-degree of freedom model for ballistic weapons, namely gun ammunition and rockets, which could compensate for helicopter downwash, projectile drag, aircraft motion, atmospheric conditions, etc **[CTE 29]**. This model was integrated with Apache's on-board fire control computer.

- The software controls and integrates multiple complex target acquisition systems. This capability to manage these systems evolved from fundamental work done at Frankford Arsenal in the mid-1970s in software algorithms. Fire control computer performance was enabled by these innovative algorithms coupled with equally innovative boresighting aids and procedures **[CTE 30]**.

- The Military Standard 1553 databus concept was a key technology required to enable the high degree of integration used by the Apache's onboard processing and sensor systems to display relevant piloting and targeting information to the crew **[CTE 31]**. The 1553 databus technology was derived primarily from work done on the Air Force's B-1 bomber program. Among the things it integrated into the fire control system was the radar frequency interferometer (RFI) **[CTE 32]**. Industry led development of the RFI. The system, which is integrated with the Longbow MMW, passively detects and analyzes radar emissions from potential targets.

- The Improved Data Modem (IDM) permits the Apache to digitally communicate crucial threat, targeting, and other operational information with other aircraft and with ground units so that the force has a unified picture of the battlefield **[CTE 33]**. This could be considered one of the first elements of network-centric warfare for Army aviation. Industry was heavily involved with development of the IDM, as was the Army's in-house avionics program.

Longbow Millimeter Wave Radar

- The AH-64D version of the Apache incorporated upgrades in several areas, but the most notable of these was the Longbow Millimeter Wave (MMW) fire control radar. Two early programs were important to advancing MMW radar applications to armed helicopters **[CTE 34]**. The first was the Frankford Arsenal/Emerson Electric Moving Target Radar System Program. The second helicopter radar of note was a Texas Instruments system that used a transmitter in the leading edge of the UH-1 Huey rotor blades and a receiver on the nose of the helicopter to detect targets. This system never left the R&D stage, but served as exploratory technology that helped to lead to the Longbow radar.

- In the 1980s, Martin Marietta defined the advantages of using a higher frequency MMW radar to detect, classify, and recognize the target and transfer the information for use by a radar seeker missile weapons system. This stimulated an Army program that (which is classified, thus limiting information on the R&D effort) yielded the Longbow **[CTE 35]**.

- Important to the MMW radar was the very fast analog circuitry provided by gallium arsenide (GaAs) semiconductors. A DOD program administered through DARPA established the industrial base to supply GaAs to the military **[CTE 36]**. This program, known as the MIMIC (microwave monolithic integrated circuits) program, provided the basis for product, process, and applications technologies.

- Another key innovation associated with the Longbow radar is its location on the mast **[CTE 37]**. A mast-mounted sight allowed the helicopter to remain concealed. Aerodynamic studies conducted at Army-NASA sites were key to the location of the Longbow radar **[CTE 38]**.

Weapons Suite

- The Apache was designed primarily as an antitank weapons system, and its main armament for that mission is the Hellfire (HELicopter Launched FIRE-and-forget) missile. The Hellfire with which the Apache was initially equipped was guided by a semi-active laser (SAL) seeker **[CTE 39]**. A new type of radar-guided Hellfire was designed for use with the Longbow **[CTE 40]**. It provided a true autonomous fire-and-forget capability.

- Below the nose of the Apache is a rapid-fire, turreted 30mm chain gun. Engineers at Hughes Helicopter came up with a new single-barreled chain gun design for the Apache that could fire 11 rounds per second, was highly reliable, resistant to dirt and wear, and could continue to fire when rounds failed **[CTE 41]**.

Modeling and Simulation and Other Enabling Technologies

Co-Located Army-NASA Research Sites

- Much of the modeling work related to helicopter development stemmed from co-location of Army engineers and scientists at NASA research facilities. For instance, interaction between the rotor and the airframe of the AH-64, particularly in the tail, caused potentially catastrophic vibrations **[CTE 42]**. Applying advanced structural dynamics analysis and comprehensive aeromechanics analysis, researchers worked with industry in a major configuration redesign of the tail, relocating the horizontal stabilizer and converting it to a movable stabilator.

- Significant advances in rotorcraft computer codes used to simulate complex helicopter aerodynamics and helicopter flight control laws resulted in development of an Engineering Design Simulator for Apache at Boeing Mesa **[CTE 43]**. This significantly reduced the time required to complete qualification and testing of the AH-64D model, and put this weapons system in the hands of the user much sooner than would have otherwise been possible.

Other Collaborative Efforts

- A recent collaborative effort that could have a far-reaching impact on many future DOD systems is the effort known as the Rotorcraft Pilot Associate (RPA) program, developed and tested by Boeing and AATD **[CTE 44]**. The RPA program developed very robust Cognitive Decision Aiding System software intended to reduce the workload of the Apache's crew in the highly complex, low altitude combat environment.

Summary of CTEs in the Development of the Stinger and Javelin Missiles

The following summary provides brief explanations of the CTEs discussed in the Stinger and Javelin *Hindsight Revisited* report. Introductory material, conclusions, discussion of any advances not dubbed CTEs, and some analysis have been omitted. This summary also dispenses with references and citations. The full report, with complete footnotes, can be found at: <http://www.ndu.edu/ctnsp/ Def_Tech/ DTP%2033%20Missiles.pdf>.

Stinger

Seeker

- The Stinger is a fire-and-forget missile, meaning that its onboard systems must be able to engage the target without further assistance from the gunner or any other external source. Stinger Basic improved upon earlier seekers by using a technique called conical scanning developed by General Dynamics **[CTE 1]**.

- The next version of Stinger, Stinger–POST (for Passive Optical Seeker Technique), made three major improvements to the seeker. The reticle seeker was replaced with a rosette scan seeker, patented by General Dynamics, which provided more accurate target discrimination **[CTE 2]**. Stinger–POST was equipped with a dual wave length detector assembly: one detector that operated at the mid-infrared and another detector that operated at the near ultraviolet (UV) **[CTE 3]**. This improved Stinger's all-aspect engagement capability—the missile was no longer as dependent on picking up engine heat, which would always be strongest at the rear of an aircraft. The third major improvement in Stinger–POST was the incorporation of integrated digital circuits to perform the seeker signal processing functions **[CTE 4]**. Conceived, developed, and demonstrated by Missile Research, Development and Engineering Center (MRDEC), the concept was embraced and incorporated by the prime contractor, General Dynamics.

- Stinger RMP introduced another major improvement. This design (as the name Reprogrammable Microprocessor indicates) enabled the onboard microprocessor to be updated with new software as new information on threats and countermeasures became available **[CTE 5]**. The reprogrammable circuitry, conceived and developed by General Dynamics, made it possible to add capability to the missile without fully redesigning it.

- The missile has an onboard battery to power the systems after launch. Based on PM-funded work at Sandia National Laboratory, a lithium-based battery was developed in the mid-1990s for use in Stinger Block I as a replacement for the initial chromate battery **[CTE 6]**.

Guidance and Control

- Though the computation necessary to track a target is straightforward, the instrumentation necessary to make it possible is complex. The Stinger is a rolling airframe missile, so the guidance system needs to know where in its roll the missile is and must angle its canards many times a second in order to achieve a consistent direction. The principle rolling airframe-related improvement over Redeye introduced with Stinger Basic was a better servomechanism to drive these canards **[CTE 7]**. The servomechanism was developed by General Dynamics. The latest version of the Stinger—Block I—has a roll frequency sensor that uses laser ring gyros to measure the roll rate **[CTE 8]**. The roll frequency sensor was developed in a collaborative effort between the PM, Raytheon, and Honeywell.

- The missile makes an adjustment in the last moments before it impacts the target, turning from the focus on the high-temperature plume of the target aircraft to the airframe itself. This is done using a Target Adaptive Guidance circuit, which was perfected for the Stinger by a collaborative effort between the contractor and MRDEC **[CTE 9]**.

Propulsion and Warhead

- Stinger propulsion performance was enhanced over earlier systems by advances in propellant technology. These were achieved principally through improvements to the propellant binder. Researchers at the Propulsion Directorate at Redstone Arsenal developed a binder polymer system based on liquid hydroxyl-terminated polybutadiene (HTPB) that enables a wider range of burn characteristics and structural capabilities than earlier binders for composite propellants **[CTE 10]**. Atlantic Research Corporation (ARC) used HTPB when it tailored the specific propellant formulations (one each for the boost and sustain phases) for Stinger. The capabilities of HTPB propellant allowed ARC to design a flight motor for Stinger in which both the boost and sustain propellant grains are bonded to the motor case **[CTE 11]**. Case-bonded propellants are cheaper, more efficient, and more reliable than earlier cartridge-loaded grain.

Modeling and Simulation

- The Stinger program was the first missile development program to utilize computer-based simulation from design through production. For Stinger, full-up simulations were developed by a community consisting of MRDEC, the prime contractor, the Army Materiel Systems Analysis Activity, the user community, and the PM office **[CTE 12]**. They simulated the results of live-fire shots and performed sensitivity experiments on the seeker on the computer. Use of the computer models saved an estimated $100M for the Stinger program.

- Between the prime contractor and MRDEC, two hardware-in-the-loop simulators were used in the Stinger's development. Basic Stinger was the first Army missile system to use a validated hardware-in-the-loop (HWIL) simulation to perform many of the system evaluation tasks previously done by extensive (and expensive) hardware testing **[CTE 13]**.

Javelin

Background

- In 1981 DARPA undertook a research program to develop an antitank missile that would make use of an imaging IR guidance system and employ a top-attack strategy, that is, the round had to be able to strike the top of a tank, where its armor is thinnest. This program was known as "Tankbreaker." The Tankbreaker program resulted in important advances in imaging IR seekers that laid the groundwork for the Javelin program **[CTE 14]**.

- A Joint Venture (JV) composed of Texas Instruments (TI) and Martin Marietta won the contract to produce the Advanced Antitank Weapons System—Medium (AAWS–M), This missile was later dubbed the Javelin. The Army decision to use a JV approach for the Javelin was an important part of the program's success **[CTE 15]**. This JV entity manages the Javelin program, but the technical work and much of the manufacturing is done by the two participating companies on the basis of a workshare agreement.

Command Launch Unit

- The imaging IR is the most vital part of the CLU. The CLU's scanning IR array is where the target recognition capability resides. The CLU developed by TI had a 240x1—later 240x2 and 240x4—focal plane array (FPA) of mercury cadmium telluride detectors, operating in the long-wave IR region of 8–12 microns **[CTE 16]**. The detectors were scanned at 30Hz in a bi-directional, interleaved fashion **[CTE 17]**. Moving from right to left the odd pixels (1, 3, 5, etc.) are sampled, then moving from left to right the even pixels are sampled. The bi-directional scan developed for the Javelin was unique and allowed a significant power savings. This scanner motor approach was later transitioned to several other TI programs.

- The CLU used a new method to normalize the detector chips. TI developed a Thermal Reference Assembly, a passive optical assembly that provides two temperature points against which each detector pixel is calibrated **[CTE 18]**. This approach was passive, requiring no additional power or control circuitry. That allowed the system designers to calibrate using existing imaging circuitry as well reduce power and save space.

- The Javelin CLU's IR detectors require cooling to a very low temperature to increase the signal-to-noise ratio. Cooling is provided by a closed-cycle Stirling engine with a cold finger projecting into a Dewar flask and up against the back of the detector **[CTE 19]**. The cooler, developed by TI with DARPA support, consumes only 1/5 watt of power yet delivers the cooling capacity required to cool the FPA in a two-and-a-half-minute period.

- Since the original design of the CLU's IR imager, progress has been made on improving the system's performance **[CTE 20]**. A DARPA-funded program in the early 1990s helped to make detectors more producible. TI also moved from four separate "through hole" circuit boards to two surface mounted boards, using advances in semi-conductor technology to save space and weight.

- The CLU's housing was originally made of aluminum. In 1999, the aluminum was replaced with a 17–layer carbon resin fiber composite [CTE 21]. This saved some weight but mostly made the housing stronger.

Seeker

- The Javelin is a fire-and-forget missile. This fire-and-forget capability came from the Joint Army/Marine Corps Source Selection Board's decision to select the JV AAWS–M design, which coupled an imaging IR system with a state-of-the-art onboard tracking system [CTE 22].

- The missile seeker for the Javelin is a two-dimensional (2D) staring FPA of 64x64 detector elements [CTE 23]. The detectors are made of an alloy of cadmium-tellurium and mercury-tellurium (termed mercury cadmium telluride or HgCdTe). Development of the 2D staring array turned out to be very difficult. TI had a manufacturing problem that risked doubling the development cost of the program and causing its cancellation. Hughes's Santa Barbara Research Center, working under a DARPA contract, developed another design for a focal plane array that could be manufactured more efficiently [CTE 24].

- The Javelin's seeker is calibrated using a "chopper" wheel [CTE 25]. The FPA is continually provided with points of reference in addition to viewing the scene. These reference points allow the FPA to reduce fixed pattern noise.

Guidance and Control

- The Javelin's tracker is the essential element of the missile's guidance and control capability **[CTE 26]**. The signals from each of the seeker's over 4,000 detector elements are passed to the FPA's readout chip. To guide the missile, the tracker locates the target in the current frame and compares this position with the aim point. If this position is off center the tracker computes a correction and passes it to the guidance system, which makes the appropriate adjustments to the control surfaces. Development of the Javelin's tracker was done by both industry and Redstone Arsenal. Texas Instruments designed and built prototypes, and Redstone provided both upgrades and an independent assessment of the tracker's capabilities. Extensive captive flight testing of the AAWS–M seekers and trackers enabled the tracker teams to test, refine, and update algorithms prior to missile firings **[CTE 27]**.

Propulsion and Warhead

- The Javelin's motor was developed by ARC, now Aerojet. ARC had adapted the design from one developed by Alliant Technology. The Javelin has an integrated launch and flight rocket motor **[CTE 28]**. Among other advantages, this integrated design kept system weight as low as possible.

- Gunner safety was a key consideration. The Javelin is equipped with a pressure release system to ensure that a malfunctioning launch motor does not cause an explosion **[CTE 29]**. The launch motor has shear pins, developed jointly by government and industry, that fracture in the event of launch motor overpressure and allow the motor to be pushed out the back of the launch tube.

- Another important propulsion design element is the burst disc that separates the launch motor and the flight motor [CTE 30]. This feature, developed by ARC, protects the flight motor from the ignition of the launch motor, yet, when sufficient pressure develops, lets the flight motor rupture the disc and send flight motor gases past it and down through the launch motor chamber.

- The Javelin missile's tandem warhead is a high explosive antitank (HEAT) round. Advances in the lethality of shaped charge rounds were made to counter the advent of explosive reactive armor (ERA). To defeat ERA, the Javelin uses two shaped charge warheads in tandem [CTE 31]. The precursor charge sets off the ERA and clears it from the path of the main charge; the main charge penetrates the target's primary armor. This concept, first applied in the Tube-launched, Optically-tracked, Wire-guided (TOW) missile, was based on work done at BRL and Picatinny Arsenal. The Javelin's designers initially struggled to make the tandem warhead work. Conventional Munitions Systems Inc. eventually contributed a successful precursor design using a two-layered molybdenum liner [CTE 32].

- The main charge had to be protected as much as possible from the explosive blast, shock, and debris caused by the impact of the front of the missile and the detonation of the precursor charge. To limit interference, a composite blast shield was developed at Redstone Arsenal and placed between the main charge and the precursor charge **[CTE 33]**.

- With the multiple warheads, variable time delay requirements, and weight and volume constraints of the Javelin and other missiles, and with safety requirements becoming more stringent, existing mechanical fusing technology was inadequate. As a result, an electronic arming and fusing effort was initiated for missile systems and applied to Javelin **[CTE 34]**. This concept, based on work done for nuclear warheads at Sandia and Los Alamos, came from engineers at Redstone Arsenal in the mid 1980s. It was given the acronym ESAF, for Electronic Safe Arming and Fire.

Modeling and Simulation

- Many of the technical achievements associated with the Javelin's development were enabled by modeling and simulation. Perhaps most importantly, the Javelin team developed an all-software simulation called the Integrated Flight Simulation (IFS) **[CTE 35]**. IFS, created by TI, does sensitivity analyses, simulates behavior of the focal plane arrays and the tracker, shows the voltages sent to the guidance unit, etc. While developed for the Javelin, the model is now also used on other missile systems.

Acronym Glossary

6.1	Basic Research
6.2	Applied Research
6.3	Advanced Development
2D	Two-Dimensional
AAH	Advanced Attack Helicopter
AATD	Aviation Applied Technology Directorate
AAWS-M	Advanced Antitank Weapons System—Medium
ABS	Agent Based Simulation
AFRL	Air Force Research Laboratory
AMC	Army Material Command
APFSDS	Armor-Piercing, Fin-Stabilized, Discarding Sabot
APG	Aberdeen Proving Ground
ARC	Atlantic Research Corporation
ARCIC	Army Capabilities Integration Center
ARL	Army Research Laboratory
ARO	Army Research Office
ASD (FMP)	Assistant Secretary of Defense Force Management Policy
ATO	Army Technology Objective
BRAC	Base Realignment and Closure
BRL	Ballistics Research Laboratory
C2	Command and Control

C4	Command, Control, Communications, and Computers
C4ISR	Command, Control, Communications, Computers, Intelligence, Surveillance, and Reconnaissance
CECOM	Communications and Electronics Command
CLU	Command Launch Unit
CSRA	Civil Service Reform Act (1979)
CTE	Critical Technology Event
CTNSP	Center for Technology and National Security Policy
DARPA	Defense Advanced Research Projects Agency
DDR&E	Director, Defense Research and Engineering
DIS	Distributed Interactive Simulation
DOD	Department of Defense
DOD69	Original *Project Hindsight* report, produced by DOD in 1969.
DOE	Department of Energy
DREN	Defense Research and Engineering Network
DSB	Defense Science Board
DU	Depleted Uranium
ECBC	Edgewood Chemical Biological Center
ENIAC	Electronic Numerical Integrator And Calculator
ERA	Explosive Reactive Armor
ESAF	Electronic Safe Arming and Fire
FBCB2	Force XXI Battle Command, Brigade and Below
FCS	Future Combat Systems
FLIR	Forward Looking Infrared

FPA	Focal Plane Array
FTE	Full Term Equivalent
GaAs	Gallium Arsenide
GE	General Electric
GIG	Global Information Grid
GPS	Global Positioning System
HCPC	Hybrid Collective Protection Equipment
HEAT	High Explosive Antitank
Hellfire	HELicopter Launched FIRE-and-forget
HgCdTe	Mercury Cadmium Telluride
HMMWV	High-Mobility Multipurpose Wheeled Vehicle
HPC	High Performance Computing
HPCMP	High Performance Computing Modernization Program
HR	Human Resources
HSI	Human-Systems Integration
HTPB	Hydroxyl-Terminated Polybutadiene
HWIL	Hardware-In-The-Loop
IDM	Improved Data Modem
IED	Improvised Explosive Devices
IFS	Integrated Flight Simulation
IHADSS	Integrated Helmet and Display Sight System
IR	Infrared
IVIS	Intervehicular Information System
JV	Joint Venture
LDP	Laboratory Demonstration Program
LQIP	Laboratory Quality Improvement Program
LSI	Lead Systems Integrator
LVC	Live-Virtual-Constructive

M&S	Modeling and Simulation
MANA	Map Aware Non-uniform Automata
MANPRINT	Manpower and Personnel Integration
MIMIC	Microwave Monolithic Integrated Circuit
MMW	Millimeter Wave
MRDEC	Missile Research, Development and Engineering Center
MRS	Muzzle Reference System
MSRC	Major Shared Resource Center
NASA	National Aeronautics and Space Administration
NBC	Nuclear, Biological, or Chemical
NBS	National Bureau of Standards
Nd: YAG	Neodymium: Yttrium Aluminum Garnet
NDAA	National Defense Authorization Act
NDU	National Defense University
NIST	National Institute of Standards and Technology
NLOS-C	Non-Line of Sight Cannon
NOE	Nap-of-Earth
NPR	National Performance Review
NRAC	Naval Research Advisory Committee
NRL	Naval Research Laboratory
NSPS	National Security Personnel System
NVL	Night Vision Laboratory
OMB	Office of Management and Budget
OPM	Office of Personnel Management
OSD	Office of the Secretary of Defense
OSTP	Office of Science and Technology Policy
PM	Program Manager
PosNav	Position/Navigation
POST	Passive Optical Seeker Technique

RDEC	Research, Development and Engineering Center
RDT&E	Research Development Test and Evaluation
RFI	Radar Frequency Interferometer
RMP	Reprogrammable Microprocessor
RPA	Rotorcraft Pilot Associate
RXD	Research or Exploratory Development
S&T	Science and Technology
SAL	Semi-Active Laser
SECDEF	Secretary of Defense
SIMNET	Simulator Networking
T&E	Testing and Evaluation
TACOM	Tank-Automotive Command
TADS/PNVS	Target Acquisition and Designation Sight and the Pilot Night Vision Sensor
TI	Texas Instruments
TOW	Tube-launched, Optically-tracked, Wire-guided
TRADOC	Training and Doctrine Command
UAV	Unmanned Aerial Vehicle
UV	Ultraviolet
WMI	Warfighter-Machine Interface

www.ingramcontent.com/pod-product-compliance
Lightning Source LLC
Chambersburg PA
CBHW051507170526
45166CB00001B/433